Object-Oriented Forth

Implementation of Data Structures

Object-Oriented Forth

Implementation of Data Structures

DICK POUNTAIN

ACADEMIC PRESS
Harcourt Brace & Company, Publishers
London □ San Diego □ New York □
Boston □ Sydney □ Tokyo □ Toronto

ACADEMIC PRESS LIMITED
24–28 Oval Road
London NW1 7DX

United States Edition published by
ACADEMIC PRESS INC.
San Diego, CA 92101

Third printing 1994

British Library Cataloguing in Publication Data

Pountain, D.
Object-oriented Forth : implementation of data structures.
1. FORTH (Computer program language)
I. Title
005.13'3 QA76.73.F24

ISBN 0-12-563570-2

Printed in Great Britain by
St Edmundsbury Press Limited, Bury St Edmunds, Suffolk

Acknowledgments

I should like to thank the following people who have helped, more or less directly, to make this book possible: Rosemary Altoft, for patience beyond the call of duty; Charles Moore, for inventing Forth; Leo Brodie, for a start in Forth; E.V.Schorre and other FIGgers too numerous to name, for further education; Alistair Mees, my mentor in matters of Forth style; Alan Winfield and Martin Worrell, for telephone support; Chuck Duff, for swapping ideas with a fellow traveller; Bill Dress, for a Heap of ideas; Larry Forsley and Thea Martin, for the Rochester Conference; Steve Pelc, for the loan of software; Bob Wallace, for producing PC-WRITE, which made writing the book a pleasure; Phil Lemmons and BYTE magazine, for keeping body and soul together during its writing; and Marion, for thousands of cups of tea. Finally, I should like to thank Professors Knuth, Hoare, Djikstra, Wirth, Tenenbaum and Augenstein whose writings taught me about responsibility in programming, and Alan Kay and his PARC colleagues, who showed me that responsibility need not exclude freedom.

Contents

Introduction

Forth is, regrettably, one of the best kept secrets in the computing world. By this I do not mean to imply that no one knows about it, but rather that few people appreciate how powerful and productive it can be when properly applied.

Since Charles Moore invented the language in the late sixties, a relatively small group of astronomers and others involved in instrument control have discovered that Forth provides the most direct, revealing and flexible way for controlling computer hardware yet invented. Furthermore, thousands of hobbyists have been introduced to the language through the admirable activities of the various branches of the Forth Interest Group (FIG). The spread of Forth in amateur circles has been aided by the fact that Forth is perhaps the only language whose compiler is sufficiently simple that keen amateurs can implement it themselves, perhaps with the help of the free listings distributed by FIG. Very often when a new low-cost personal computer is introduced to the market, the first language (other than its inevitable ROM Basic) to be put up on it is FIG-Forth.

Unfortunately Forth has also acquired a bad reputation in large sections of the professional programming community, and it is to this that my "best kept secret" remark refers. This reputation is not entirely without justification. Forth provides a degree of freedom in programming style which is unmatched by any other language. Combined with the activities of thousands of self-taught enthusiasts, this has resulted in the phenomenon which unbelievers derisively call "write-only programming"; the development of highly cryptic and personalized programming styles which can make Forth programs utterly unreadable by outsiders.

Another reason for the relative unpopularity of Forth in professional circles is the misperception of its role as a language. Unlike fully compiled languages such as Fortran, Pascal and C, Forth provides an integrated, interactive programming environment which does not lend itself well to use as a production language for "packaged" commercial software. It is very difficult for example to link Forth programs to modules written in other languages, and many of the cheaper Forth systems still do not provide facilities for writing stand-alone programs to run under industry-standard operating systems. Forth enthusiasts have perhaps even contributed to this misperception, by occasionally extravagant claims and a reluctance to accept the importance of industry standards.

But those critics who like to claim that no well known commercial program has ever been written in Forth (which isn't quite true anyway), are missing the point. Forth's true strength is as a language for those people who have to write their *own* programs; not as a sausage-machine for producing commercial software. Forth provides the quickest and easiest way to probe the corners of an unfamiliar hardware system that has so far been devised. Engineers and scientists who are constantly faced with having to write control programs for novel hardware have traditionally been in the vanguard of Forth use. A glance at the delegates list for the increasingly successful annual Rochester Forth Conference will reveal names from every major research laboratory (both military and civil), and most of the major hardware manufacturers in the USA with quite a few from Europe too.

The key to Forth's success as a "rapid prototyping" language lies in its unique combination of interactivity, unrestricted low-level access to the hardware, and the unlimited flexibility to extend the language itself. It allows its user to do anything that could be done in machine language, instantly and interactively, using trial and error techniques. The only other popular high-level language which offers equivalent levels of machine access is C, and a good Forth programmer will have finished and gone home while the C programmer is still wading through the "edit, compile, link, load, run, crash, edit, recompile" cycle.

This book is about making Forth more generally useful than it already is; it is an attempt to break out of the engineering and hardware development ghetto in which Forth has so far prospered.

Part of the attraction of Forth lies in its extensibility, which allows programmers to use it as a "construction set" rather than a rigid set of rules. As a consequence, standard Forth provides no way to deal with structured data; its data objects are 16-bit integer values. It does however provide a uniquely powerful mechanism for *building* data structures, through the use of the words CREATE and DOES>. The full exploitation of these words is not easy, and tends to be one of last things that a Forth programmer comes to grips with. This is partly because they provide a facility not found in other languages, namely to change the syntax of the language itself and partly because they demand the ability to clearly distinguish between compile-time and run-time, and to write programs which act in both these time domains.

Forth also permits the programmer to modify the action of the compiler itself by writing compiling words, a facility provided in few other languages, and only feasible because the Forth compiler is so elegantly simple that a programmer can actually understand its operation.

By using both these techniques, the experienced Forth programmer can build any data structure, of any complexity, that he or she wishes, from a simple array to the most complex tree structure. However it is likely that every such programmer will choose a different way to accomplish this, and so the myth of "write-only code" is given substance. The fact that such programs are doing things that are impossible, or only available to professional compiler writers, in other languages will not mollify the critics.

But there is another issue at stake, beyond the readability of programs. It is now widely recognized that the *reusability* of code is crucial to efficient software production. This view has developed from the simple libraries of Fortran sub-routines (some of them decades old) which are still in use, through the "software tools" philosophy so eloquently propounded by Kernighan and Plauger, up to the point where modern languages like Modula-2 and Ada are entirely designed around the concept of sealed and reusable modules.

Reusability is perhaps too weak a word for what I am talking about here. It is certainly a great advantage not to have to constantly re-invent the wheel. But the advantages go beyond the mere labour saved. Once a reusable module has been thoroughly debugged and tested, then it can be incorporated into new programs in the confidence that it *works*. The time saved in debugging is much greater even than the time saved in recoding tasks that someone else has already done.

This book sets out to develop *systematic* ways of constructing complex data structures in Forth. Rather than every Forth programmer tackling the problem in a different and individualistic fashion, I shall suggest mechanisms that allow any kind of data structure to be built using a few easy to use syntactic extensions to the language. My hope is that these mechanisms strike an acceptable balance between the freedom which is what attracts us to Forth in the first place, and a discipline which can make Forth a more suitable language for data intensive programming than it presently is.

This is not an introductory Forth book. To understand it you will need a firm grounding in the basics of Forth programming. Anyone who has diligently worked through one the many excellent introductory Forth books now available, such as Leo Brodie's "Starting Forth" (Prentice-Hall 1981) or Alan Winfield's "Complete Forth" (Sigma 1983), should have no trouble at all in following it. Keen beginners in Forth might be able to cope by reading it in *conjunction* with such an introductory book. There is nothing at all to stop you *using* the techniques in this book without understanding the internal workings of the code that implements them. To do that you only need to type in the code and compile it. But I suspect that few real Forth enthusiasts will be happy using code that they do not understand back, front and sideways.

My hope is that some Forth implementers will read this book and find the suggested extensions sufficiently useful to incorporate into their products. I have found the use of object-oriented programming techniques as described in Chapter 2 quite exhilarating and productive, and I cherish the hope that they may have the same effect on others. However it would be folly to expect all these others to have to understand how, say, TYPE>...ENDTYPE> works in order to use it. This book is therefore aimed as much at Forth implementers as users. The marvellous thing about the Forth community is that the distinction between these two categories is so blurred, and I welcome any Forth user at all who might find the book useful.

The book is designed above all to be read *while sitting in front of a computer*. While you will gain some information from reading the text alone, there is no substitute for trying it out on the spot. Some of the more difficult concepts expounded, such as second order defining words, will become very easy once you have typed in the code and played with a few examples. This is what Forth programming is all about, and it should be what reading a Forth book is all about. Since Forth is incrementally compiled, you do not even need to perform a marathon typing task; each piece of code can be typed in as it is encountered, and given a thorough thrashing until you understand what it does.

All the code in all the chapters of book, if compiled, would not occupy more than about 4K of memory, so this technique of reading-and-doing is open to users of anything between a Sinclair Spectrum and a VAX. Before beginning though, please take the time to read the next section on Standards, to discover whether or not you need to make any changes to your Forth system before proceeding.

Standards and Notation

All the Forth code in this book is written to conform as nearly as possible to the most recent standard, Forth-83. However system-modifying code such as this needs to have access to the individual fields of Forth word headers (name field, link field, code field and parameter field or body). Direct addressing of these fields is forbidden to Standard programs, and so in that sense this book does not contain standard programs. Worse still, the '83 Standard does not include all the words necessary to address these fields (it only has >BODY to convert a code field address into a parameter field address). I have adopted a set of field addressing words which are "in the spirit" of Forth-83, and are implemented in Laboratory Microsystems Inc.'s compilers (one of which, PC-Forth v 2.0, I used in writing the book). They originate from an experimental Standard proposal by Kim Harris, which is likely to be incorporated into the next standard.

The words and their actions are as follows :

```
>BODY    -- convert code field address to parameter field address [Standard]
>NAME    -- convert code field address to name field address
>LINK    -- convert code field address to link field address

BODY>    -- convert parameter field address to code field address
NAME>    -- convert name field address to code field address
LINK>    -- convert link field address to code field address

N>LINK   -- convert name field address to link field address
L>NAME   -- convert link field address to name field address
```

There are some words which I use frequently for readability which are not included in the standard, and will be defined here. TRUE and FALSE are two boolean constants to be used as values for flags. They are defined by :

```
0 CONSTANT FALSE     FALSE NOT  CONSTANT TRUE
```

LATEST returns the name field address of the most recently compiled word in the dictionary, and can be defined by :

```
: LATEST    CONTEXT @ @ ;          ( --- addr )
```

Verbal comments in this book are delimited by the -- symbol (as used in Ada), which marks everything to the end of the line as a comment, rather than by the normal Forth brackets. This was done for purely aesthetic reasons, as in heavily commented code the preponderance of brackets on the printed page becomes unattractive. The word can be defined as follows :

```
: --      BLK @ IF     >IN @  64 / 1+ 64 *
                ELSE  #TIB @
                ENDIF >IN ! ;               IMMEDIATE
```

Most readers will not wish to type in the comments in any case, and those that do may prefer to use the normal brackets. I have retained brackets for stack notation diagrams, to distinguish these from other comments.

Finally, I detest the use of THEN in the IF...ELSE...THEN control structure so vehemently that I refuse to publish code containing it. I consider it to be ugly and counter-intuitive, and to symbolize all that gives Forth a bad name in the wider computing community. So here is the definition of ENDIF, as used throughout this book :

```
: ENDIF    [COMPILE] THEN ;    IMMEDIATE
```

Those who disagree with my little obsession are at liberty to replace it by THEN.

Some other non-standard words will be required (particularly those concerned with the creation of string constants) but these will be explained and defined at the appropriate points in the text.

Translation to Forth-79

Some care has been taken to make translation to the older Forth-79 standard as straightforward as possible. In fact much of the early work for the book was done on a Forth-79 system and translated later on.

The only changes needed to convert examples from the book into Forth-79 are to replace the header field access words >BODY etc., and to alter the use of ' ("tick") and PICK.

The translation of the header field addressing words is complicated by two facts. Firstly the 79 Standard contained no header field addressing words at all. However most serious implementations borrowed the word set from FIG-Forth, namely PFA,CFA,LFA and NFA. The second complication is that these words all use the parameter field address (PFA) as a reference point, whereas the Forth-83 words use the code field address :

```
NFA   -- convert parameter field address to name field address
CFA   -- convert parameter field address to code field address
LFA   -- convert parameter field address to link field address

PFA   -- convert name field address to parameter field address
```

Here are translations for those combinations of field addressing words that I have used in this book :

```
BODY> >LINK       translates into     LFA

BODY> >NAME       translates into     NFA

N>LINK            translates into     PFA LFA

' >BODY           translates into     [COMPILE] '
```

>NAME occurring on its own cannot be translated directly. The definition will have to be reorganized so that a parameter field instead of a code field address is supplied. This is usually easy to do because ' in Forth-79 produces a parameter field address rather than a code field address. This is reflected in the last translation above, where the >BODY is simply omitted. The [COMPILE] is necessary because ' was an immediate word in Forth-79.

What is to be done for readers whose Forth systems do not have either of these sets of field address conversion words? In order to use the code in Chapter 2 of this book you will have to write them. How they are defined depends on the finer details of your Forth implementation, which you will have to discover (by poring over hex dumps if absolutely necessary). For typical Forth systems they are in fact defined very simply, merely consisting of the addition or subtraction of a small integer. For example >BODY and BODY> are almost always :

```
: >BODY 2+ ;    : BODY> 2- ;
```

For less orthodox systems (e.g. with separated headers or complex dictionary linking), you will need intimate knowledge of the implementation, and I cannot guarantee that my code will work anyway. The only remaining translation problem concerns PICK. I have deliberately refrained from using PICK as far as possible, and in fact it only occurs once (in Chapter 4, in the definition of ADJUST.HANDLES). Its argument needs to be increased by 1 for Forth-79 (e.g. 2 PICK becomes 3 PICK).

I have not tried any of the code in this book on a pure FIG-Forth system, and so cannot guarantee that it will work.

Stack Annotation

One of the most important aspects of commenting Forth programs is the annotation of the stack effect of the individual words. Without such annotation it is difficult or impossible for another person to use the word, let alone to modify it. A convention has arisen among Forth programmers for annotating stack effects. A small "diagram" shows the content of the stack before the word is executed, followed by some dashes as a separator, followed by the contents of the stack after the word has executed. For example the stack effects of the word DUP can be annotated as (n --- n n), i.e. it takes one single number from the stack and leaves two single numbers.

Because the code contained in this book deals with complex data structures, I have had to extend the notation used in stack diagrams. In addition to the normal use of **n** to mean a single number, **d** for a double number, **char**, **addr**, **flag** and the other standard abbreviations, I have had to invent notations for the addresses of various data structures, because a simple **n** or even **addr** does not convey enough information. These have been made as nearly self explanatory as possible, and are always explained in the text before use. For example the annotation (list --- node) should be read as "the address of a list structure is taken from the stack, and the address of a list node is left on the stack". The symbol **?** is used to show that the stack contents at that point could be anything at all. It is encountered frequently in the chapter on abstract data types, where the stack contents after executing some words depend upon what operation was executed. To take a simple example, the stack effect after

```
: TEST   CODE.ADDRESS @ EXECUTE ;
```

has been executed cannot be depicted in a diagram; it depends upon what the code whose address is stored in CODE.ADDRESS actually does.

If a definition has no stack annotation, this can be taken to imply that it has no stack effect.

One further convention has been adopted. When a word takes text from the input stream at run-time (because its definition contains WORD, ' , or CREATE), then this is signified by changing the three dashes which separate "stack before" from "stack after" into three plus signs. So (+++ flag) would mean that the word takes nothing from the stack, but takes a following word from the input stream, and it leaves a boolean flag on the stack.

Some of the code in this book defines second order defining words; words which define words which define words. Annotating the stack effects of such words is extremely difficult, as there is a time dimension involved. To give as much help as possible to the reader, I annotate both the CREATE and the DOES> parts of words which use this construction.

Naming Conventions

No elaborate naming conventions have been adopted for this book. The names of words have simply been chosen to be as explanatory as possible, helped by the use of a Forth system which allows long names. Readers may wish to shorten some of the names in the interests of quicker typing; my concern has been for readability always.

Although I personally prefer to use lower case names for my own words, to distinguish them from Forth primitive words, I have adhered to upper case throughout this book, partly because it is traditional in Forth literature, and also because it helps the printers with their unenviable task of distinguishing Forth words from the rest of the text.

The nearest thing to a naming convention lies in the use of multi-word names. I have used a full-stop symbol to break up such words when they are for internal use only (e.g. IN.TYPE.DEF?, MAKE.INSTANCE), whereas I use a hyphen in words that are intended for the user (eg. DEFINES-TYPE, ARRAY-OF). Words that return a boolean flag have been given names that end in ?.

1 Records

Forth is at first sight a low-level, untyped language. Most current Forth implementations are 16-bit, by which we mean that the stack, so important to Forth, is 16 bits wide, and the addresses compiled into the threaded code are also 16-bit (32-bit Forth's are beginning to emerge with the new generation of microprocessors, but there is no standard yet and they will not be dealt with here). Therefore the only data types standard Forth recognizes are the 8-bit byte, the 16-bit single number and the 32-bit double number. Moreover both the byte and the double number are "underprivileged" types in the sense that they occupy less or more than one stack cell and they are often manipulated by 16-bit operators (e.g. there is no DROP or SWAP operation for bytes).

However, Forth is unlike most other languages in that it allows the user to extend and modify the compiler itself. It is perfectly possible to write extensions to handle extra data types such as quadruple (64-bit) numbers, or character strings.

In a similar way, standard Forth provides very few data structures, but instead includes a powerful mechanism for building them as required. This is quite in keeping with the Forth philosophy of keeping the basic kernel as small as possible.

The standard Forth data structures consist of just 16-bit scalar VARIABLEs and CONSTANTs, and classical Forth teaching suggests that even these be sparingly used, the stack to be used for data storage whenever possible.

In practice most programmers soon discover a need for more structured data objects, and it is very common to find, at the least, single dimensioned arrays implemented as extensions in commercially available Forth systems.

The Forth mechanism for building new data structures employs the pair of words CREATE and DOES>. These two words permit the user to define new defining words (words which are themselves used to define new words, just as :, VARIABLE and CONSTANT are).

In this chapter we shall run through some simple examples of the use of CREATE...DOES>, and then see how to use them to create simple structured data types.

Create and Does

The word CREATE simply creates a new Forth dictionary header for the name which follows it in the input stream. So

```
CREATE FRED
```

produces a dictionary entry called FRED. This dictionary entry has a name field, link field and code field but its body (or parameter field) is quite empty. CREATE advances the dictionary pointer so that HERE lies immediately after the new header, and thus points to its empty parameter field.

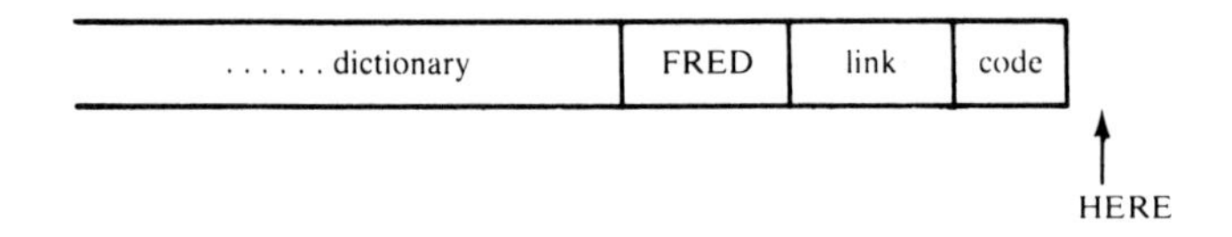

When a word made by CREATE is executed, its action is merely to put its parameter field address onto the stack. So executing FRED will leave the address of its empty body on the stack.

When employed in the definition of a new defining word, CREATE enables the defining word to create dictionary entries.

DOES> is used to modify the *behaviour when executed* of the new words that CREATE produces. All such words have by default the behaviour just mentioned, namely that they put their parameter field address onto the stack. If they are required to do more than this (and they usually are), then this required action is placed after DOES> in the definition of the defining word.

The Forth defining word CONSTANT could be defined as follows :

```
: CONSTANT    CREATE ,          ( n +++  )
              DOES>  @   ;      (  --- n )
```

Note that the stack effect of CREATE...DOES> definitions needs to be documented twice; the CREATE part should record the stack effect when the defining word is executed, i.e. at compile time, while the DOES> part should show the stack effect of words defined by this defining word, i.e. the run-time effect. Let us follow what happens when CONSTANT is executed, as in :

```
23 CONSTANT FRED
```

Firstly CREATE grabs the name FRED from the input stream, and makes a dictionary header for FRED. Then , is executed which takes 23 from the stack and compiles it into the dictionary at HERE. Since HERE coincides with FRED's parameter field, the effect is to fill the parameter field with 23.

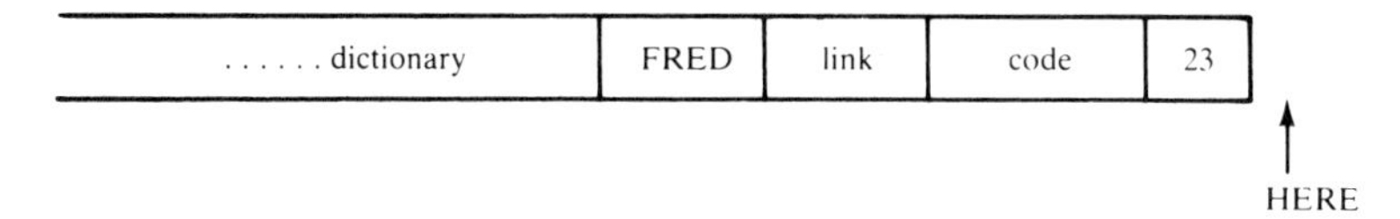

When FRED is executed, its parameter field address is placed on the stack. However there is now a further action. The DOES> part of CONSTANT's definition says that @ is also to be executed by the defined word. This takes the parameter field address from the stack and fetches its contents, namely 23, giving us the required behaviour for a Forth CONSTANT.

The action of CREATE...DOES> can be summarized like this :

```
: new.defining.word
                    CREATE   code to be executed when
                             new.defining.word is executed
                    DOES>    code to be executed by words defined with
                             new.defining.word
;
```

Arrays

Before moving on to more advanced applications, let us see how CREATE...DOES> can be used to implement a single dimensional array data structure. The defining word we shall create is called []VARIABLE, and it requires the size of the defined array to be placed on the stack thus :

```
10 []VARIABLE  new.array
```

The elements of **new.array** can be accessed by placing an index on the stack, so :

```
  6 new.array @   -- fetches the contents of element 6 of new.array)
5 7 new.array !   -- stores 5 into element 7 of new.array)
```

A possible definition of []VARIABLE is :

```
: []VARIABLE                 ( n +++   )
             CREATE            -- make new header
               2*              -- convert size to bytes
               ALLOT           -- allot space for array body
             DOES>           ( n --- addr )
               SWAP            -- swap PFA and index
               2*              -- convert index to a byte offset
               +   ;           -- index + PFA = element's address
```

When an array name defined by []VARIABLE is executed it returns the address of the required element on the stack, which can then be treated like an ordinary variable using ! and @.

This definition could be improved in many ways. For example, the indices for the elements of new.array run from 0 to 9 but some people might find it more natural and less error prone to use instead indices 1 to 10.

Furthermore []VARIABLE does no checking of any sort. For a useable version we would want at least :

(i) A check (using DEPTH) that a size has been actually been placed on the stack.

(ii) A check that this size is positive (a negative size would corrupt the dictionary).

In addition we could provide a run-time (i.e. DOES> part) check that the supplied index lies within the declared bounds of the array, though many Forth programmers will prefer to forgo this security in the interest of efficiency.

Here is a version incorporating some improvements :

```
: []VARIABLE    DEPTH 0=
                IF ." No size supplied!"
                ELSE  DUP 0< IF ." Negative array size!"
                             ELSE CREATE  2*  ALLOT
                                  DOES>   SWAP 1- 2*  +
                             ENDIF
                ENDIF ;
```

It is straightforward to extend this technique to multi-dimensional arrays, though the syntax becomes clumsy and error prone when several indices are required on the stack in addition to the data to be stored. It's a classic Forth demonstration, used in many introductory Forth books, to show how arrays can be made "active" by putting suitable code in the DOES> part. For instance one could create an array type which computes the running average of all its elements and stores it in element 0.

Structured Data Types

Many modern programming languages such as Pascal, C, Modula 2 and Ada provide facilities for the user to define complex data structures by combining the simple data types provided by the language. Such structured data types are a powerful aid to the programmer when dealing repeatedly with groups of data items of similar structure.

For example, when designing a database management program it is natural to group together named fields (possibly of different types) into records:

```
NAME     24 bytes
ADDR1    40 bytes
ADDR2    40 bytes
TEL      16 bytes
```

It is very convenient to be able to create a single named data structure which could contain the whole of this conglomeration and allow it to be manipulated as a single object when required, but still allow access to the individual named fields.

In Pascal such an object is called a *record,* while in C it would be called a *structure.* Forth does not provide any equivalent, but we can easily construct one with the aid of CREATE and DOES>.

Records

Let us begin by outlining what it is we wish to achieve. We want to be able to construct defining words which define complex *named* structures whose individual parts are also accessible by name. It would be useful if we could arrive at a syntax which allows such defining words to be created in a way which is simple, readable and doesn't expose too much low-level detail of the implementation.

As a first attempt, let us define a record type called **address.record** which reflects the structure of the database example above, i.e. four fields of 24,40,40 and 16 characters. This is a possible approach :

```
0   CONSTANT name                      -- byte offsets into the record
24  CONSTANT addr1
64  CONSTANT addr2
104 CONSTANT tel

: ADDRESS-RECORD   CREATE 120 ALLOT    -- allocate 24 + 40 + 40 + 16 bytes
                   DOES> +  ;          ( n --- addr )
                                       -- add field offset to get address
```

Then we can declare a new record with

```
ADDRESS-RECORD  JohnDoe
```

and access its individual fields by, for example,

```
addr1 JohnDoe COUNT TYPE
```

(which assumes that a string has been stored into **addr1** with its length in the first byte).

Although this works after a fashion, it is unsatisfactory for several reasons.

(i) The calculation of the offsets and the total memory requirement for ALLOT is left up to the programmer, providing plentiful opportunities for error.
(ii) The record **JohnDoe** considered as a whole object must be referred to by either using **name JohnDoe** (which implies the special knowledge that **name** is its first field), **0 JohnDoe** which is ugly and confusing or **' JohnDoe** which is perhaps the least objectionable way. The latter method will return the code-field address in Forth-83, and so we need to do >BODY to get the address of the actual data storage area.
(iii) Too much low-level implementation detail is visible, which detracts from the comprehensibility of the code.
(iv) The RPN syntax (field before record-name) is not conducive to readability.

As a second attempt we might try :

```
: .name   ;
: .addr1  24 + ;
: .addr2  64 + ;
: .tel    104 + ;

: ADDRESS-RECORD    CREATE 120 ALLOT ;
```

This produces a syntax closer to the "dot notation" used in Pascal and other languages :

```
JohnDoe .addr1   COUNT TYPE
```

It also overcomes objection (ii) in that the address of the whole record is now returned by plain **JohnDoe**. More importantly, it suggests a further step which will attack objections (i) and (iii).

By using CREATE...DOES> to define the field-names, we can both automate the offset calculation and hide some of the nuts-and-bolts detail :

```
VARIABLE  TOTAL.BYTES      0 TOTAL.BYTES !

: FIELD                             ( n +++  )
          CREATE TOTAL.BYTES @      -- total so far is the offset
                 ,                  -- store it in the field name
                 TOTAL.BYTES +!     -- now bump the total
          DOES>  @ +  ;              ( addr --- addr)
                                    -- get offset and add to record addr
```

The word FIELD accumulates the total size of the record and stores the individual field offsets into the bodies of the field-names. Field-names when executed add their offset to the record start address, which must be on the stack, and so leave the starting address of the field on the stack. We could now declare, by analogy with the previous version :

```
24 FIELD .name
40 FIELD .addr1
40 FIELD .addr2
16 FIELD .tel

: ADDRESS-RECORD   CREATE TOTAL.BYTES @ ALLOT  ;
```

This is very much better as the programmer no longer has to keep track of the field offsets manually. It suffers from an intolerable drawback however. Only one type of record can be defined in this way; if we defined a second kind (say PAYROLL-RECORD) then the value held in TOTAL.BYTES would be changed, and ADDRESS-RECORD would no longer work correctly. To avoid this we need to store the value from TOTAL.BYTES in ADDRESS-RECORD itself, thus treating TOTAL.BYTES as a purely temporary storage place :

```
24 FIELD .name
40 FIELD .addr1
40 FIELD .addr2
16 FIELD .tel

: ADDRESS-RECORD  [ TOTAL.BYTES @ ] LITERAL
                  CREATE ALLOT ;
```

This now puts the correct value onto the stack when ADDRESS-RECORD is used to define a record, regardless of what the current value of TOTAL.BYTES is. Indeed we could dispense with TOTAL.BYTES altogether and redefine FIELD so that it uses the stack to transmit the total size to ADDRESS-RECORD. But to do so would make the improvements we shall introduce later harder to read and understand.

The definition of ADDRESS-RECORD has now been generalized to the point where its body will be the same for any record definition. If we define, for example,

```
2 FIELD .real
2 FIELD .imag

: COMPLEX   [ TOTAL.BYTES @ ] LITERAL  CREATE ALLOT ;
```

then the definition of COMPLEX is identical to that of ADDRESS-RECORD. This suggests that it would be useful to factor out this code so that it doesn't need to be repeated every time. However we cannot simply take it out into a colon definition because the brackets cause compile time actions and the value of TOTAL.BYTES would be compiled at the wrong time.

A more radical solution is to use instead a *second order defining word*, i.e. a word which defines words which define words, to create our records. As we shall see later this step also opens up new vistas of power and flexibility.

Second order words can be defined by using a CREATE...DOES> whose DOES> part itself contains a CREATE...DOES> construct. This is not a step to take lightly because it can become difficult to envisage exactly what is going on, and when! If you are not yet completely comfortable with the way CREATE...DOES> works, go back over the previous material until you are.

Rather than compiling the value of TOTAL.BYTES directly using LITERAL, we shall store it into the record defining word itself, making the latter behave rather like a constant :

```
: DEFINES-TYPE      CREATE  TOTAL.BYTES @ ,       -- store size
                    DOES>   @ CREATE ALLOT ;
```

A record definition may now be written like this :

```
24 FIELD .name
40 FIELD .addr1
40 FIELD .addr2
16 FIELD .tel

DEFINES-TYPE  address.record
```

This latest version successfully overcomes all four of the objections raised above, and leads to a syntax which very clearly expresses what is being done without revealing any distracting detail.

The word DEFINES-TYPE creates the record defining word in the dictionary, replacing the colon definition we were using before. Colon was rather too general for our needs and showed too much of the nuts-and-bolts detail which we wish to hide; DEFINES-TYPE is a customized defining word which only creates the sort of object we are interested in. The record defining word **address.record**, when executed, puts the size of a record instance onto the stack, then CREATEs an instance of that size. An *instance* merely means a particular example of a type of record. When we say

```
address.record  JohnDoe
```

JohnDoe becomes an instance of the type **address.record**.(It's unfortunate that the word TYPE is already used in Forth to print strings; I hope that it will not introduce too much confusion if we also use it in this wider but unrelated sense, to mean a kind of record.)

Here is the code collected together for reference. It will make life simpler later on if we factor DEFINES-TYPE into two parts, calling the instance creation part MAKE.INSTANCE :

```
VARIABLE   TOTAL.BYTES        0 TOTAL.BYTES !

: FIELD    CREATE                     ( n +++ )
             TOTAL.BYTES @             -- total so far is the offset
             ,                         -- store it in the field name
             TOTAL.BYTES +!            -- now bump the total
           DOES>                      ( addr ---  addr)
             @ +    ;                  -- perform address calculation

: MAKE.INSTANCE    CREATE ALLOT  ;          ( n +++ )

: DEFINES-TYPE     CREATE                   ( +++ )
                     TOTAL.BYTES @           -- total is instance size
                     ,                       -- store size
                     0 TOTAL.BYTES !         -- reset the total
                   DOES>  @ MAKE.INSTANCE ;
```

Record Size

It would be convenient for record instances to have knowledge of their own total size, so that they can be easily moved around in memory using CMOVE. This is easily accomplished by storing the total size in the first two bytes of the record, and adjusting all the offsets accordingly. An instance of **address.record** would then look like this diagrammatically :

size

120	name	addr1	addr2	tel

This in turn requires that the size be stored in the record defining word so that it is available during instance creation :

```
VARIABLE   TOTAL.BYTES        2 TOTAL.BYTES !

: FIELD    CREATE                     ( n +++  )
             TOTAL.BYTES @ ,           -- total so far is the offset
             TOTAL.BYTES +!            -- now bump the total
           DOES>  @ +   ;              ( addr --- addr )
                                       -- perform address calculation

: MAKE.INSTANCE    CREATE               ( n +++ )
                     DUP  ,             -- store instance size
                     ALLOT   ;          -- allocate fields

: DEFINES-TYPE     CREATE                       ( +++ )
                     TOTAL.BYTES @ ,             -- store instance size
                     2 TOTAL.BYTES !             -- allow for size field
                   DOES>  @  MAKE.INSTANCE ;
```

Now the data in an instance of **address.record** such as **JohnDoe** can be easily copied into a new record without the programmer having to remember or look up its size :

```
address.record JohnDoe   address.record temp
. . . . . . . . . . . . .

JohnDoe temp OVER @  CMOVE
```

This process could be taken further so that every field in the record stored its length in its first byte (limiting field lengths to 256 bytes) or two bytes. This of course would involve a considerable memory overhead, which grows worse the more fields a record has :

120	24	name	40	addr1	40	addr2	16	tel

An alternative is to store the length of fields in the **field-names** themselves, and change their run-time behaviour so that the count as well as the address is returned, ready for use with CMOVE or FILL. This solution is especially attractive for fields which store strings, for which an address and count will be required for all operations :

```
: STRINGFIELD   CREATE                ( n +++  )
                  TOTAL.BYTES @ ,     -- total so far is the offset
                  DUP ,               -- store field length from stack
                  TOTAL.BYTES +!      -- ...and bump the total
                DOES>                 ( addr --- addr count)
                  2@                  -- get length and offset
                  ROT +               -- add offset to base address
                  SWAP  ;             -- put count on top
```

Here are two sample operations on such a field :

```
JohnDoe .name  32 FILL       -- initialize name field to all spaces
PAD  JohnDoe .name  CMOVE    -- move a string from PAD into name field
```

It is much less attractive for numeric fields where the count is of no relevance and will need to be dropped; this spoils the transparency of the syntax and increases opportunities for error :

```
JohnDoe .salary DROP @
```

It may sometimes be worth distinguishing the two types of field and using both NUMFIELD (identical to our original FIELD definition) and STRINGFIELD, at the price of having two types of field-name with different behaviours.

Nested Records

Records and structures in Pascal and C may be nested; in other words, a field in a record may itself be a record. This is a powerful feature, which encourages the programmer to decompose very complex data structures in a hierarchical fashion. For example, we might have :

```
personnel.record =  address.record
                    payroll.record
                    medical.record
                    pension.record
                    promotion.record
```

This nesting ability is already within our grasp, since record defining words contain the length of their instances. To extract this length it is necessary to use 'tick' and to obtain the parameter field address of the defining word. A nested record structure could be defined by, say,

```
                          2  FIELD .patientnumber
' address.record >BODY  @  FIELD .address
' medical.record >BODY  @  FIELD .medical

DEFINES-TYPE  patient.record
```

To access the **name** field of an instance of **patient.record** it is now necessary to use a *double* field-name reference, first referring to the name of the embedded record and then the field within that record :

```
patient.record  JoeBlow
JoeBlow .address .name  COUNT TYPE
```

The syntax can be cleaned up somewhat by defining a word called, for example, USE

```
: USE  ' >BODY @ ;
```

so that we can say

```
                    2  FIELD .patientnumber
USE  address.record FIELD .address
USE  medical.record FIELD .medical

DEFINES-TYPE  patient.record
```

An instance of **patient.record** now looks like this :

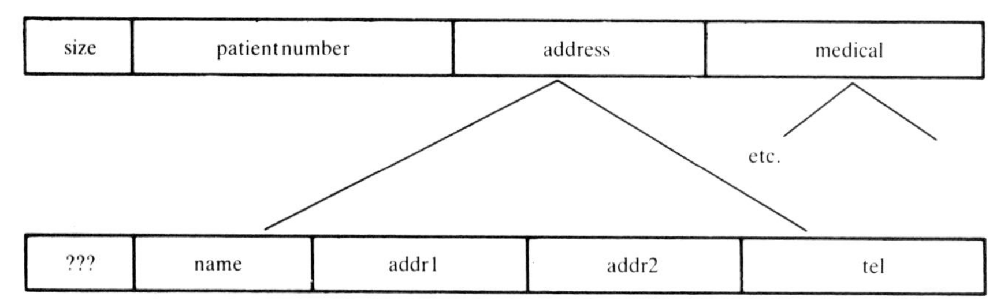

Such nesting may be carried out to any depth, but it may become difficult to keep track of all the field-names and to avoid field-name clashes if very deep nesting is used.

There is however one serious deficiency with this system, which the alert reader may have spotted. As it stands, the system permits access to the fields of embedded records, but does not allow their size to be found; hence it is not possible to copy such a sub-record independently of the whole. This is because the size field of an "embedded" record such as **.address** does not get filled in with its correct size, but contains a garbage value. The size value extracted by USE only goes to bump up the total space allocation stored in TOTAL.BYTES, and MAKE.INSTANCE only writes a single size field describing the whole **patient.record**.

In order for MAKE.INSTANCE to write the correct values into the size fields of such embedded records, we should have to preserve much more information than is done at present. In fact we would need to keep a "map" inside **patient.record** which showed the offset of each embedded record and its size; MAKE.INSTANCE would then use this map to allot space and write correct size fields. The complication introduced would be very considerable.

This provides a powerful argument for our alternative mechanism (see STRINGFIELD above) in which sizes are stored in the field-names themselves. In this case the data fields of a record are "clean", that is they contain nothing but data. This removes the need for any "mapping" to be done when new instances are created, at the cost of using a few DROPs to dispose of the size when it is not required. Our previous example would then look like this :

```
patient.record  JoeBlow
JoeBlow .address DROP .name  COUNT TYPE
```

Note that this solution still leaves a wasted two byte field at the beginning of every embedded record, since all records (as opposed to fields) still contain a size field. However the value in this field is no longer needed.

With Blocks

When a program is doing a lot of work with multiple fields of a particular record, it can soon become verbose and irritating to have to always prefix the field-name with the record-name, as in :

```
: PRINTL   COUNT TYPE CR ;

JohnDoe .name  PRINTL  JohnDoe .addr1 PRINTL JohnDoe .addr2 PRINTL
JohnDoe .tel   PRINTL
```

Pascal provides a neat shorthand in its WITH construct. At the beginning of a block the programmer may say **WITH JohnDoe DO**. Throughout that block the field-names alone may be used to access the fields; it becomes implicit that **JohnDoe** is the record being referred to. With a little ingenuity (and at a slight loss of run-time efficiency) we can add an equivalent feature by writing a new version of FIELD :

```
VARIABLE CURRENT.RECORD        0 CURRENT.RECORD !

:  -{     CURRENT.RECORD !   ;            ( addr ---     )
:  }-     0 CURRENT.RECORD ! ;

: FIELD    CREATE                          ( n +++  )
             TOTAL.BYTES @ ,
             TOTAL.BYTES +!
           DOES>                           ( addr --- addr )
             CURRENT.RECORD @ ?DUP         -- is a default set?
             IF  SWAP  ENDIF               -- then use it
             @ +  ;
```

The above example could now be written more concisely as :

```
JohnDoe -{ .name PRINTL .addr1 PRINTL .addr2 PRINTL .tel PRINTL }-
```

The trick works equally well with nested records, where a partial "path-name" could be made the default record :

```
JoeBlow .address -{ .name PRINTL .addr1 PRINTL .addr2 PRINTL .tel PRINTL }-
```

This scheme is not entirely satisfactory as it does not allow the default record to be overidden should a *different* record need to be accessed inside a "with" block, nor does it allow nested blocks. A fuller solution which supports nested blocks can be produced, using a stack to preserve the previous default record, but is probably too complicated to be worthwhile in this case.

Binding Time and Efficiency

Though the above scheme works quite well and achieves the goals we originally set out, it sacrifices a certain amount of run-time speed, when compared to equivalent programs using ordinary Forth variables. This is because whenever we wish to access the contents of a field at run-time, a fetch and an addition have to be performed to calculate the field address. The execution of this code is an overhead that would not be present if we merely used simple Forth variables to represent the fields. In fact this overhead is not necessary, and in eliminating it we shall discover a principle of some importance.

The implementation of record structures devised above uses what computer scientists call "late binding". All this means is that we are leaving it until run-time (when a record is actually being used) to determine which particular instance of a record type is being addressed. Our field-names take a record's address, at run-time, and add the appropriate offset to access the required field, "binding" the field-name to a value. If we were only interested in using records in interpreted mode from the keyboard (as in the examples above), then this would be a reasonable and indeed the only way to proceed. But what if we use records inside other colon defined Forth words, which is much more likely? Using the record type COMPLEX defined above, we might produce a definition like :

```
COMPLEX x
: initialize    0  x .real !  0  x .imag ! ;
```

Here it is known at the time that **initialize** is compiled, that the instance of COMPLEX referred to is **x**. So it is quite wasteful to wait until **initialize** is executed to compute the addresses of the fields **.real** and **.imag** inside **x**. If instead we compute them at *compile-time*, then we can compile the actual addresses and there will be no run-time speed overhead at all. **initialize** will run just as fast as if we had instead used :

```
VARIABLE xreal     VARIABLE ximag
: initialize    0 xreal !  0 ximag ! ;
```

This is called "early binding"; performing the binding of names to values at compile-time instead of run-time. In the case of our records it is well worth pursuing, because it means that we have bought the beneficial effects of a structured data type at no cost in execution speed at all (though at a very small cost in compilation speed, which is only incurred once).

The modifications required to incorporate early binding are quite simple. Record and field-names need to become IMMEDIATE words so that they are executed at compile-time. And instead of leaving the field address on the stack, they must compile it into the enclosing definition. Let us illustrate using the simplest version of records for clarity :

```
VARIABLE  TOTAL.BYTES         0 TOTAL.BYTES !

: FIELD   CREATE TOTAL.BYTES @ ,        -- total so far is the offset
                 TOTAL.BYTES +!         -- now bump up the total
                 IMMEDIATE              -- make field-name immediate
          DOES>  @ +                    -- perform address calculation
                 [COMPILE] LITERAL ;    -- and compile into definition

: MAKE.INSTANCE   CREATE ALLOT IMMEDIATE ;   -- instance name immediate

: DEFINES-TYPE    CREATE  0 TOTAL.BYTES !
                  DOES>  @ MAKE.INSTANCE ;
```

Notice that MAKE.INSTANCE needs to be IMMEDIATE too, because we want record instance names to "go off" at compile-time and put their base address on the stack.

[COMPILE] LITERAL just compiles the address of LITERAL (which is an immediate word and would otherwise be executed) into the run-time code for a field-name. When the field-name itself is executed, LITERAL is executed with the calculated field address on the stack, and so compiles this value as a literal into the definition inside which the immediate field-name was executed. [COMPILE] LITERAL is therefore a way of deferring the forced compilation of a value; it is only of use when defining an immediate word, and we shall see many more examples of its use later on. Spend a little time making sure you understand how it works, as it can be rather brain-hurting when first encountered.

Let us recap a little by defining COMPLEX in full :

```
2 FIELD .real
2 FIELD .imag
DEFINES-TYPE COMPLEX

COMPLEX x

: initialize    0  x .real !  0  x .imag ! ;
```

Now when **initialize** is compiled **x**, **.real** and **.imag** are all executed at compile-time. The address calculation is performed and the actual address of the relevant field is compiled into the definition, just as if a single Forth variable had been used in place of the record-name/field-name combination.

But what happens if we execute a field reference directly from the keyboard in interpretive mode; say **x .real** @ ? In the simple case we're discussing here it works just fine. The fact that the words are immediate is of no consequence when they are executed interpretively. The address calculation @ + is executed normally leaving the result on the stack, and [COMPILE] LITERAL apparently does nothing. In fact it does something invisible and relatively harmless; it compiles the address of LITERAL into the dictionary at HERE, so consuming 2 bytes of work-space. However we cannot rely on this harmless behaviour in the general case. In the next chapter we shall develop early-binding schemes in which the compilation behaviour is much more complex and will not work correctly if interpreted.

As a result the words need to be "state-smart", that is they need to know whether they are being interpreted or compiled, and to do different things accordingly. We can make FIELD state-smart quite cheaply :

```
: FIELD    CREATE TOTAL.BYTES @ ,          -- total so far is the offset
                  TOTAL.BYTES +!           -- now bump up the total
                  IMMEDIATE                -- make field name immediate
           DOES>  @ +                      -- perform address calculation
                  STATE @ IF                    -- if compiling....
                      [COMPILE] LITERAL    -- ..then compile address
                  ENDIF ;
```

Recall that the variable STATE is set to 1 when Forth is compiling, and 0 when interpreting. This new version does not waste any dictionary space through unwanted LITERALs when used interpretively, at the cost of a slight penalty in interpretation or compilation speed (due to the IF test). The efficiency of the compiled code is exactly the same as in the previous version.

This is a trade-off we shall use again and again in the next chapter; interpretation speed against efficiency of compiled code. It is a worthwhile trade-off because records used interpretively can never be time critical, and they will usually be used in this way for debugging purposes only. It is not possible to perform looping in interpretive mode in standard Forth, and so no interpreted record can ever be accessed more than once. This being so, a penalty of a few microseconds is hardly a disaster. The same argument applies to the one-off penalty incurred at compile-time. In an inner loop in a compiled word however, such a penalty matters very much, and it is to removing this penalty that early-binding schemes are dedicated.

Nesting and Early Binding

This extra efficiency from early binding does not come completely free. Unfortunately nested records will not now work when compiled into a colon- definition. The reason is simply that because field-names now have an immediate action at compile-time, multiple field-names no longer work correctly. Taking the example used previously :

```
                    2  FIELD .patientnumber
USE  address.record  FIELD .address
USE  medical.record  FIELD .medical

DEFINES-TYPE  patient.record


patient.record  JoeBlow
```

When applied interpretively this still works :

```
JoeBlow .address .name  COUNT TYPE
```

.address and **.name** merely add their respective offsets to the base address on the stack. But if we try to do the same thing inside a definition, trouble appears :

```
: TEST   JoeBlow .address .name  COUNT TYPE  ;
```

This will fail to compile with a "definition not finished" error message. The problem is that **.address** executes at compile-time and *compiles* the address of **JoeBlow**'s address field into the definition, leaving nothing on the stack for **.name** to use when it in turn executes at compile-time.

There is an easy solution to this dilemma, albeit one which spoils the syntax a little. We can permit the early binding of **JoeBlow .address** but *defer* the binding of **.name** until run-time, thus sacrificing part of the efficiency gained. This is accomplished as follows :

```
: TEST   JoeBlow .address [COMPILE] .name  COUNT TYPE  ;
```

By forcing the compilation of **.name**, we defer its execution till run-time, when the required address will be on the stack. If we had a deeply nested record, *every* field-name bar the first would need to be prefixed with [COMPILE].

Clearly this is not an ideal solution. It would be much better if all the earlier field-names were to compute addresses at compile-time, but leave them on the stack instead of compiling them, only the final one compiling its result into the definition; in other words the exact reverse of what we have just achieved. Such a scheme can be made to work using yet another new version of FIELD in which a flag variable (similar in action to CURRENT.RECORD) switches on the compiling action :

```
VARIABLE  BIND.NOW    FALSE BIND.NOW !

: FIELD   CREATE TOTAL.BYTES @ ,         -- total so far is the offset
                 TOTAL.BYTES +!          -- now bump up the total
                 IMMEDIATE               -- make field name immediate
          DOES>  @ +                      -- perform address calculation
                 STATE @ BIND.NOW @ AND    -- if compiling AND ready
                 IF                        -- to bind.......
                    [COMPILE] LITERAL      -- ..then compile address
                 ENDIF
                 FALSE BIND.NOW ! ;

: \     TRUE BIND.NOW ! ;     IMMEDIATE    -- set the binding flag
```

Now we could say, for example,

```
: TEST   JoeBlow .address \ .name  COUNT TYPE  ;
```

The \ word, which instigates binding, will always precede the last field-name in a chain, and the result is full early binding again, with the absolute address of a field being compiled into a definition.

This solution illustrates a trick which we shall be using again in the next chapter, namely the introduction of new *states* into the Forth compiler. Standard Forth is a four-state system; it is either compiling or interpreting (flagged by STATE) from either the keyboard or from a virtual memory buffer (flagged by BLK). However it can sometimes be very convenient to create new special purpose states, such as "binding or not binding", to imbue words with more flexible behaviour.

Records and Disk I/O

Most applications in which record structures are useful will involve storing data on disk. All the examples used so far in this chapter have been suggestive of a database application, in which one would naturally expect the data to be stored on disk and brought into memory as required.

Unfortunately disk I/O is one of the least attractive aspects of Forth, and as implemented in standard systems it is quite unsuited for inputting and outputting records.

Standard Forth performs disk I/O using unstructured "blocks" of a fixed, 1024-byte size. These blocks are mapped directly onto physical disk sectors, and so standard Forth acts as a disk operating system as well as a programming language. Disk data are addressed by taking a byte offset into a memory buffer containing such a block, read into memory by the word BLOCK, whose action is to take a block number and return a buffer address.

Any program for inputting and outputting records will therefore be grossly inefficient unless the size of the record happens to be an exact divisor or multiple of 1024 bytes. For example, suppose we write an application which requires a disk file of records, each 350 bytes in size. Only two such records can be fitted into a block, as the remaining space will then be only 324 bytes.

It is not easily possible to let a record run over a block boundary, because the way Forth allocates buffer space in its virtual memory system cannot guarantee that consecutive blocks are loaded into contiguous memory. Furthermore, the addressing of such records would become indecently complicated. We would therefore be left with little choice but to waste 324 bytes out of every 1024 bytes of disk space.

These reflections prompt a small digression into a controversial area. Forth programmers are divided between "purists" who regard Forth as being a stand-alone operating system as well as a programming language, and "revisionists" who regard Forth as merely a programming language (like Pascal or C) which should run under the native operating system of the host computer.

The author sits firmly in the revisionist camp, having first learned Forth using a dialect which runs under the CP/M operating system, and prepared this book using a dialect running under PC-DOS. Such dialects have full access to the services provided by the underlying operating system, which invariably include the ability to read and write sequential and random files of arbitrarily sized records. These systems usually also provide the standard Forth block structure for compatibility with standard programs, though the blocks are contained in files constructed by the host operating system.

Unfortunately, since standard Forth is meant to stand alone, there is no standard for interfacing Forth to host operating systems, and each implementation tends to do it in a different way. It is therefore not possible to give an example of an efficient record-based application without tying it to a specific dialect and operating system. This is particularly irritating given that device-independent I/O is featured by most other modern programming languages such as C.

Following these rather bilious observations, let us see what can be done about record I/O in standard Forth.

We have seen earlier that it is easy to move the data storage portion of a record in memory, using CMOVE. It is therefore straightforward to move record data into a disk buffer and hence write it to disk. All that remains is to devise a system for addressing such records on disk, as follows :

```
VARIABLE REC.SIZE  VARIABLE RECS.PER.BLOCK

-- Declare type of record held by a file

: FILETYPE  ' >BODY @ REC.SIZE !           -- get record size
            1024 REC.SIZE @ /              -- compute number of records that..
            RECS.PER.BLOCK !  ;            -- ...fit into a block.

-- Get the address in disk buffer of record 'recnum'       ( recnum --- addr)

: FINDREC   RECS.PER.BLOCK @ /MOD 1+       -- block number and offset in records
            SWAP 1-                        -- adjust offset to start from 1.
            REC.SIZE @  *  SWAP            -- turn into byte offset.
            BLOCK +  ;                     -- load block and add in offset.

-- Put a record to disk file

: PUT     FINDREC REC.SIZE @ CMOVE UPDATE ;               ( rec recnum ---  )

-- Get a record from disk file

: GET     FINDREC SWAP REC.SIZE @ CMOVE ;                 ( rec recnum ---  )
```

To use this code, it is first necessary to create a new blocks file, which is usually just a matter of erasing the default file FORTH.SCR or FORTH.BLK from the disk, whereupon Forth will create a new empty one. If your system lets you create a named file, so much the better.

Then a record type needs to be declared. Let us use COMPLEX from our previous examples :

```
2 FIELD .real
2 FIELD .imag
DEFINES-TYPE COMPLEX

COMPLEX Z
```

Now declare that the file is to contain records of type COMPLEX :

```
FILETYPE COMPLEX
```

Record data can now be written to or read from the file as follows :

```
Z 23 PUT     -- stores contents of Z as 23rd record on file.
Z 56 GET     -- gets 56th record from file into Z.
```

Note that only the data fields of a record are stored on disk, and not the Forth header. Records are now used being used as structured program variables. Note also that this system will only work for record sizes less than 1024 bytes, and that it may waste disk space as outlined above.

The system provides random access via the record number, but it is potentially a profligate user of disk space, as it stores records "sparsely", i.e. it makes no attempt to compact the file to fill gaps. For example, if we were to store just two records with numbers 1 and 10000, then the file would occupy 58k of space! However it is easy to write a database application which simulates sequential access, by maintaining a pointer to the last record stored and using this pointer value as the record number. The file pointer would be incremented by one every time a new record was written. Of course this file pointer needs to be preserved on disk (so that the application can always ascertain the file size), for which purpose a special first record could be used.

Recap

Here is the final version of the records code collected together, complete with all the "bells and whistles" :

```
-- working variables and flags.

  VARIABLE TOTAL.BYTES            2 TOTAL.BYTES !
  VARIABLE CURRENT.RECORD         0 CURRENT.RECORD !
  VARIABLE BIND.NOW           FALSE BIND.NOW !

-- delimit a block in which a default record is in force.

  : -{      CURRENT.RECORD !    ;  IMMEDIATE     ( addr ---    )
  : }-      0 CURRENT.RECORD ! ;   IMMEDIATE

-- declare a field name.

  : FIELD   CREATE                              ( n +++  )
               TOTAL.BYTES @ ,                  -- store offset
               TOTAL.BYTES +!                   -- bump offset count
               IMMEDIATE
            DOES>                               ( addr --- addr )
               CURRENT.RECORD @ ?DUP            -- is a default set?
               IF  SWAP  ENDIF                  -- then use it
               @ +                              -- compute field address
               STATE @ BIND.NOW @ AND           -- if compiling AND ready..
               IF [COMPILE] LITERAL ENDIF       -- ...bind early
               FALSE BIND.NOW ! ;               -- reset binding state

-- make an instance of a record type (internal use only).

  : MAKE.INSTANCE   CREATE                      ( n +++  )
                       DUP  ,                   -- store instance size
                       ALLOT                    -- allocate fields
                       IMMEDIATE ;

-- create the record defining word.

  : DEFINES-TYPE    CREATE                      ( +++ )
                       TOTAL.BYTES @ ,          -- store instance size
                       2 TOTAL.BYTES !          -- reset the count
                    DOES>  @ MAKE.INSTANCE ;

-- cause binding to occur; placed before last field name in nested record.

  : \               TRUE BIND.NOW ! ;  IMMEDIATE


-- permit a previously defined record type to be used as a field.

  : USE      ' >BODY @ ;


-- File I/O words.

-- working variables

  VARIABLE REC.SIZE
  VARIABLE RECS.PER.BLOCK

-- Declare type of record held by a file

 : FILETYPE  ' >BODY @ REC.SIZE !           -- get record size
             1024 REC.SIZE @ /              -- compute number of records that..
             RECS.PER.BLOCK !  ;            -- ...fit into a block.

-- Get the address in disk buffer of record 'recnum'      ( recnum --- addr)

 : FINDREC   RECS.PER.BLOCK @ /MOD 1+       -- block number and offset in records
             SWAP 1-                        -- adjust offset to start from 1.
             REC.SIZE @  *  SWAP            -- turn into byte offset.
             BLOCK +  ;                     -- load block and add in offset.

-- Put a record to disk file

 : PUT      FINDREC REC.SIZE @ CMOVE UPDATE ;               ( rec recnum ---  )

-- Get a record from disk file

 : GET      FINDREC SWAP REC.SIZE @ CMOVE ;                 ( rec recnum ---  )
```

This is a small amount of code to implement such a powerful facility, though it still lacks several important features compared to Pascal or C. One such feature is the *discriminated union* or variant record, the addition of which would make a good exercise for the interested reader (a variant record is one with two or more alternative structures which can be selected depending on the value of a field; for example 'spouses name' might be included or excluded from a record according to the value of 'marriage status').

This brevity is a testament to that unique property of Forth which lets us modify the compiler itself in a very easy fashion. The action of this code is not so easy to understand however, which is not really surprising. Most other language compilers are far too complicated for a user to even consider modifying them; this level of programming would normally be reserved for professional compiler writers alone.

I am not suggesting that this style of programming is necessary (or even desirable) when writing routine Forth applications. However the *result* of this code is a data-structuring facility with a clean and comprehensible syntax which makes writing certain applications much easier. Whether or not you consider the effort worthwhile will depend upon the sort of applications you write, and how well standard Forth supports you in writing them.

The real purpose of this chapter however has been to introduce a number of important programming ideas which will be needed in the next chapter to introduce a far more powerful and useful facility, namely object oriented programming using abstract data types. These techniques are :

(i) The use of second order defining words, built by nesting CREATE...DOES>.

(ii) The concept of binding time, and the use of immediate words to force early binding.

(iii) The concept of deferring compilation of values with [COMPILE] LITERAL.

(iv) The introduction of new state variables and "state smart" words.

Please be sure that you understand how all these techniques work, if necessary by re-reading this chapter, otherwise the next chapter will be very hard going.

2 Abstract Data Types

We have seen in the last chapter how structured data types, analogous to those provided in Pascal or C, can be implemented in Forth with relative ease. Little was said about the operations (i.e. the words) that would be used to manipulate records and fields within records. It was assumed, however, that these would be conventional Forth words in the dictionary, and as such globally accessible to the programmer and to other Forth words.

Records offer a great convenience to the programmer by encapsulating a number of pieces of data and allowing them to be manipulated as a single entity; programs which use such complex data structures can be made shorter, cleaner and more readable by the use of records. However records as implemented here do not offer anything in the way of additional security. Since field-names are just global Forth definitions in the dictionary, there is nothing to prevent the programmer using a field name from one record type to reference a record of a different type. The result would of course be garbage if fetching data and corruption of the other fields if storing data (except in the unlikely event that the field offsets just happen to be the same).

Abstract data types provide a way to combine the convenience of structured data with added security. An abstract data type specifies not only the structure of the data to be manipulated, but also the operations that can be performed on such data. Moreover, in an abstract data type, the actual physical representation of the data is hidden from the programs which use it, and data can *only* be accessed by using the prescribed operations.

As an illustration, let us suppose that we have defined an abstract data type called STACK, which has the allowed operations EMPTY? (a test to see if the stack is empty), PUSH, and POP. An instance of STACK, say MYSTACK, can be created in the same way that we created instances of COMPLEX in the last chapter. But MYSTACK can only be used via the words EMPTY?, PUSH and POP. No other operations at all are permitted. The way MYSTACK is constructed is not visible; there are no field-names which give us the addresses of the different parts of MYSTACK. Perhaps it is implemented as an array, or then again as a linked list. All we are allowed to know, or need to know, is that a stack is something you can push things on to, or pop things off (and that it might become empty).

Abstract data types provide a number of very important benefits to the programmer:

(i) They isolate the effects of changes in a program. If we were to write a program which uses instances of STACK only in the prescribed way, such a program would be unaffected by any alteration to the way STACK is implemented. Say it became desirable to change from an array to a linked-list implementation of STACK; only the operations EMPTY?, POP and PUSH would need be rewritten, and the rest of the program would work just as before. This sort of modularity is already strongly encouraged by Forth, but it is not enforced. A conventional Forth implementation would allow the possibility of accessing a stack by, say, direct manipulation of the stack pointer, which then ties the program to a particular physical representation of the stack.

(ii) They provide security. Restriction of the operations allowed on a type means that many program errors will be caught at compile-time, which would otherwise have survived as run-time bugs. The Achilles' heel of our Record implementation, namely that field-names of the wrong type may be used in error, would be eradicated.

(iii) They encourage powerful program design techniques. A problem may be decomposed into sub-problems by considering both the data structures required to represent entities, *and* the operations to be performed on these entities. This style of thinking gets much closer to the way the real world works than do previous computing methodologies. Instances of an abstract data type can be thought of as *objects*, which have both *attributes* (the data) and *behaviour* (the operations), and this style of programming is often called *object-oriented* programming.

Languages which support object-oriented programming to a greater or lesser degree include Simula-67 ("classes"), Modula 2 ("modules"), Ada ("packages"), CLU ("clusters") and Smalltalk-80 ("classes" and "objects"). Apple Corp. has developed an object-oriented dialect of Pascal called Clascal, for developing software on the Macintosh and several dialects of object-oriented C (e.g. C++) are now available.

Forth programming already has a flavour of object-orientation about it because its subprogram units, i.e. words, can contain data as well as code, and they are independently executable. Furthermore Forth already incorporates a mechanism for information hiding in the shape of its vocabulary mechanism.

To make our record types into abstract data types, it is necessary to hide the definition of the record structure, and the operators which manipulate it, from the rest of Forth. A type becomes a "black box" whose internal details are of no concern to the application programmer who will use the type. The organization should be such that it becomes impossible to use an operator word on an object of the wrong type; that is, a type-checking mechanism is required.

Encapsulation

In the Forth context, encapsulation of a structure merely means that it does not appear in the dictionary, and so cannot be found in a dictionary search. Either it gets executed by an indirect reference from a word which does appear in the dictionary, or else there must be a way to make it visible when required.

The vocabulary mechanism could be used for this latter purpose. If each record type definition were placed into its own vocabulary, then it would be possible to use the same field and operation names in different types, and to guarantee that only the correct version will be found :

```
VOCABULARY ADDREC IMMEDIATE
ADDREC DEFINITIONS

24 FIELD .name
40 FIELD .addr1
40 FIELD .addr2
16 FIELD .tel

DEFINES-TYPE  address.record

: name.is    ASCII . WORD DUP C@  24 MIN      -- get the string; truncate?
             .name  SWAP CMOVE  ;             -- move into field

: name.show  .name COUNT TYPE ;               -- print name field

etc..........
```

The price paid for the security gained is a very unwieldy syntax, for the name of the vocabulary will have to be used as an additional prefix when any field or operation names are used :

```
ADDREC address.record JohnDoe  FORTH
..........
ADDREC JohnDoe name.is Mr John Frederick Doe.  FORTH
```

Moreover the security provided is of a poor quality because the encapsulation must be turned on and off manually by the programmer; forgetting to return to FORTH will leave the ADDREC vocabulary wide open, and so nothing has been gained.

A much more powerful mechanism would result if the record-names "knew" which vocabularies they belonged to, and could automatically open and close them when invoked. This can be arranged, using surprisingly little code.

Two tricks are required. Firstly we will build a substitute encapsulation mechanism to replace vocabularies. This can be achieved by juggling with link field contents at compile-time to create private dictionaries which are invisible to the normal Forth search order (Schorre,1980). Secondly we will create record instances which contain within them a "key" which unlocks this private dictionary when the record-name is executed.

Private Dictionaries

The concept can perhaps be made clearer by a couple of diagrams. Let us represent the normal Forth dictionary structure so :

← A ← B ← C ← D ← E ← F ← G

The words A to G are linked together, with G's link field pointing to F's name field and so on. When Forth is trying to find a word to execute or compile, it traverses this linked list as it searches.

A private dictionary mechanism can be built like this :

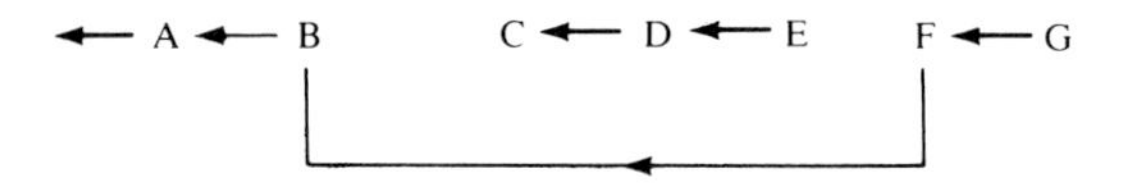

By redirecting the link field of F to point to B's name field, we have sealed up C, D and E so that they cannot be found during a dictionary search. To Forth they effectively do not exist. At first glance this seems pretty useless, as C, D and E cannot see the rest of the dictionary either, which means that their definitions must be very limited indeed (they would have to be empty or refer only to each other!). But this is to ignore the time element. What if the sealing process were to take place *after* C, D and E have been compiled, i.e. at the time that F itself is compiled. Then C, D and E can "see" the whole dictionary during their compilation and so can be definitions like any other :

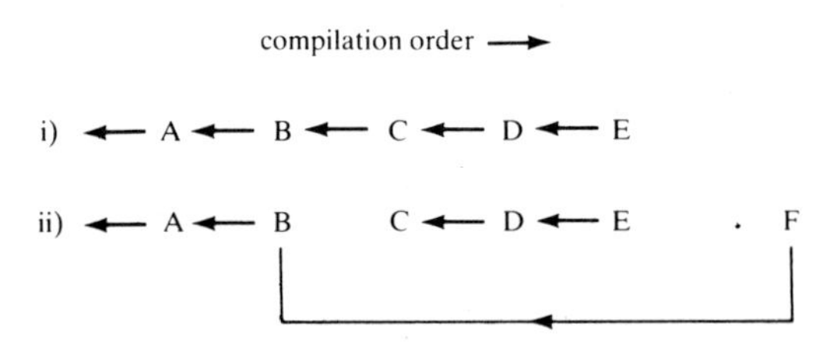

compilation of F triggers link rearrangement

Encapsulation can be achieved by placing execution-only words in the source code which do their work at compile-time. Here are two very simple definitions which will have almost the desired effect :

```
: TYPE>       LATEST  ;                -- put NFA of latest definition on stack

: ENDTYPE>    CREATE HERE               -- create a header, store stacktop in
                     BODY> >LINK ! ;    -- ...its link field
```

These would be used as follows in the source code :

```
: A  etc......... ;
: B  etc......... ;

TYPE>                    -- puts NFA of B on stack

: C  etc......... ;      ---+
: D  etc......... ;         +--  these definitions become private
: E  etc......... ;      ---+

ENDTYPE> F               -- seals the "module"

: G  etc......... ;
```

TYPE> puts B's NFA on the stack when it is executed. ENDTYPE> compiles a header for the new word called F and links it to B, creating the situation depicted in this diagram :

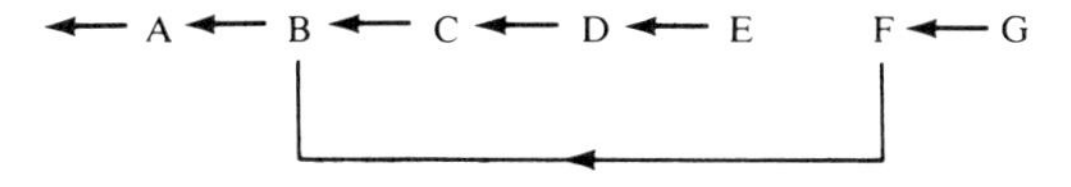

A dictionary listing would show only G F B A ....

This simple system is not sufficient though. The words need to store more information, which can then be used by "duly authorized" words to gain entrance to the "private dictionary" area of C,D and E.

Using the same letters A,B,C etc. to identify stack items, here is the code for a more useful version. In practice it turns out to be more convenient to use variables rather than the stack to transmit information from TYPE> to ENDTYPE>; otherwise we should have to be sure that nothing happened between TYPE> and ENDTYPE> which altered the stack contents. Later on we shall have even more information to transmit, and since arbitrary amounts of arbitrary code may lie between TYPE> and ENDTYPE>, it is preferable not to have to worry about maintaining the integrity of the stack.

```
VARIABLE PUBLIC    VARIABLE LASTPRIVATE

: TYPE>      LATEST PUBLIC !              -- NFA of next public word B
             CREATE                       -- make header
             HERE LASTPRIVATE ! ;         -- PFA of last private word C

: ENDTYPE>   LATEST                       -- NFA of first private word E
             CREATE                       -- make header
             HERE BODY> >LINK
                  PUBLIC @ SWAP ! ,       -- seal private dictionary
             0 LASTPRIVATE @
               BODY> >LINK !  ;           -- break link to main dictionary
```

TYPE> now creates a dictionary header and passes its LFA to ENDTYPE>. This allows the private dictionary to be sealed off from the main dictionary; the word C is made a "stopper" at the end of the private dictionary, by putting a 0 into its link field, which forces Forth to stop its search there. In other words, unlike most standard Forth vocabularies, these private dictionaries do *not* chain back to the main FORTH vocabulary. This is essential to the purpose, which is to prevent objects from responding to any operation other than those defined in their private dictionary; it is also another reason why the standard Forth vocabulary mechanism will not do the job.

```
: A  etc......... ;

: B  etc......... ;

TYPE> TYPENAME

: C  etc......... ;        ---+
: D  etc......... ;           +--  operations for type TYPENAME
: E  etc......... ;        ---+

ENDTYPE> TYPENAME

: G  etc......... ;
```

E's NFA is stored into TYPENAME (which is in the normal dictionary) so that we can retrieve it and use it as a "key" to get into the private dictionary.

Lock and Unlock

Having invented a means of encapsulating the definition of an abstract data type, it remains to invent a means by which its operations can be accessed upon request by an object that belongs to the type.

To do this we shall use the words UNLOCK and LOCK, which switch the context vocabulary to and from the private dictionary. UNLOCK will require that the "key", that is the NFA of the first word in the private dictionary, be placed upon the stack. The following simple definitions do the trick :

```
VARIABLE STASH

: UNLOCK     CONTEXT @   DUP @   STASH !   !   ;         ( key ---    )

: LOCK       STASH @   CONTEXT @   !   ;
```

UNLOCK takes the key from the stack and stores it in the location pointed to by the contents of CONTEXT, but only after it has preserved the original contents of this location in the variable STASH. CONTEXT is a standard Forth user variable whose contents point to the vocabulary which is to be used for dictionary searches. CONTEXT @ returns the address of the vocabulary, and CONTEXT @ @ returns the NFA of the first word in this vocabulary. The effect of UNLOCK is therefore to make the private dictionary into the context vocabulary; since this dictionary is not linked to Forth, the words it contains are the *only* words which can be executed after UNLOCK.

LOCK simply reverses this process and restores the previous context vocabulary again.

The required behaviour for an abstract data type can be had by making objects of that type perform UNLOCK, then look up and execute the required operation, and finally perform LOCK. This can be done in the following way :

```
: DO.OPERATION    BL WORD SWAP                              ( key +++     )
                  UNLOCK FIND LOCK
                  IF    EXECUTE
                  ELSE ." unrecognized operation" ABORT
                  ENDIF  ;
```

DO.OPERATION takes the following word from the input stream, looks it up in the private dictionary and executes it if found, otherwise issuing an error message and ABORTing.

Notice that LOCK is performed immediately after FIND and *before* EXECUTE. This is an important security measure. Consider the case where an error condition leading to an ABORT or QUIT arises during the execution of the operation. If LOCK were to be placed after EXECUTE, it would not be executed in such a case. This would leave a Forth system in which only the private dictionary words were available, and all the Forth words necessary to rectify the situation were locked out; a re-boot would be required after every minor error during program development. With the above coding, the only possibility of such a disaster is if the FIND fails in a way which leads to ABORT. In Forth-83 systems this should not be possible.

Note that to adapt this word to Forth-79, you will have to take account of the different action of FIND. The BL WORD will not be necessary as Forth-79 FIND takes a word from the input stream, not a string address from the stack. However in some Forth-79 systems, an ABORT may occur if FIND encounters an empty input stream, so typing an object name without a following operation name will produce the fatal error just mentioned. The only way to make such a system fully secure is to redefine ABORT, if possible, so that it performs the LOCK operation (see also later section THE OBJECT STACK).

Instance Variables

Only one mechanism is now missing in order for an abstract data type to be defined, namely a way of referring to the components of a data object. The FIELD mechanism developed in the last chapter will work in this role, though we shall see later that it can be considerably improved.

It is more natural to think of the components of a data object as being described by *instance variables* rather than the *fields* of a record and so the name FIELD will be changed to VAR to reflect this change of viewpoint. The names of the components of an abstract data type are not visible to the programmer outside the type definition, and may only be used in defining the type's operations. For example, if we define a COMPLEX data type :

```
TYPE> COMPLEX
2 VAR REAL
2 VAR IMAG
... operations
ENDTYPE> COMPLEX
```

then the operations are performed on the "variables" REAL and IMAG. These are not in fact variables at all, because rather than representing storage spaces, they represent *offsets* into the storage space of the *instances* of COMPLEX which will be created in the future. Nevertheless when we write the code for operations, these names can be treated as referring to the actual storage fields in any instance of COMPLEX. This illustrates the true meaning of "abstraction"; an instance variable name is an abstract description which can be realized in any number of instances of the type.

In the interests of brevity, TOTAL.BYTES has also been renamed SIZE. The code for VAR can be written, by analogy with the last chapter, as :

```
: VAR    CREATE   SIZE @ 2+ ,                      ( n +++  )
                  SIZE +!
                  IMMEDIATE
         DOES>    @ [COMPILE] LITERAL
                  COMPILE + ;
```

The 2+ is required to adjust the offset for the presence of the key, which occupies the first two bytes of every object. Note that this is not a truly early binding definition. It compiles the variable's offset as a "delayed" literal, and then the code to add it to the object address, which must be on the stack at run-time. Full early binding would demand that the addition be performed at compile-time, and the resulting data field address be compiled, as with our records in the last chapter. It should be clear though that this is intrinsically impossible for an abstract data type. The base address of an instance can *never* be known at the time the operations are compiled, because they are compiled before any instances exist; abstraction implies a distancing of the operations from their application.

Nevertheless some modest gain in speed (about 20%) can had by compiling the offset as a literal, rather than merely doing @ + at run-time. We are safe in taking this step because instance variables can *only* be compiled into operation definitions, and can never be used in interpretive mode.

A full implementation can now be produced, by writing suitable versions of MAKE.INSTANCE and ENDTYPE> which produce the correct object behaviour :

```
VARIABLE SIZE           VARIABLE PUBLIC
VARIABLE LASTPRIVATE    VARIABLE STASH

: TYPE>      LATEST PUBLIC !            -- NFA of first public word
             CREATE                     -- make header
             HERE LASTPRIVATE !         -- PFA of this new word
             0 SIZE !  ;

: VAR    CREATE   SIZE @ 2+ ,                      ( n +++  )
                  SIZE +!
                  IMMEDIATE
         DOES>    @ [COMPILE] LITERAL              ( addr ---  )
                  COMPILE + ;
```

```
: UNLOCK     CONTEXT @   DUP @   STASH !   !   ;              ( key ---    )

: LOCK       STASH @   CONTEXT @   !   ;

: DO.OPERATION    BL WORD SWAP                        ( addr key +++ ??)
                  UNLOCK FIND LOCK
                  IF    EXECUTE
                  ELSE ." unrecognized operation" ABORT
                  ENDIF  ;

: MAKE.INSTANCE    CREATE                     -- make a named instance
                     DUP @ ,                  -- store key into instance
                     2+ @ ALLOT               -- allot storage space
                   DOES> DUP @                -- put PFA and key on stack
                         DO.OPERATION ;

: SEAL       HERE BODY> >LINK
             PUBLIC @ SWAP !                  -- seal private dictionary
             0 LASTPRIVATE @
             BODY> >LINK !    ;               -- break link to main dictionary

: ENDTYPE>   LATEST                           -- NFA of first private word
             CREATE SEAL                      -- make typeword and seal
               , SIZE @ ,                     -- store key and storage size
             DOES> MAKE.INSTANCE ;
```

Notice that ENDTYPE>, like DEFINES-TYPE from the last chapter, is a second order defining word.

Let us follow through an example of the use of these words to define a type called COMPLEX. The syntax looks like this :

```
TYPE> COMPLEX
2 VAR REAL
2 VAR IMAG
: !!  DUP IMAG ROT SWAP !  REAL SWAP ! ;   -- store a complex value
: @@  DUP REAL @  SWAP IMAG @ ;              -- fetch a complex value
ENDTYPE> COMPLEX
```

When this is compiled, only the word COMPLEX appears in the dictionary. An instance of complex can be created by

```
COMPLEX X
```

and X will only respond to the two operations we have defined for COMPLEX :

```
23 17 X !!          ok

X @@   . .   17 23    ok

X +    unrecognized operator
```

The internal structure of the word COMPLEX looks like this :

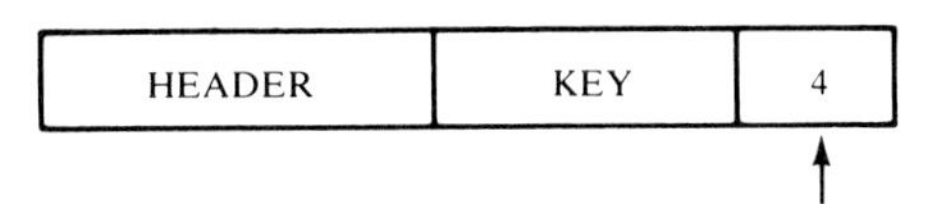

size of storage required by instances

while the instance of COMPLEX, X, looks like this :

HEADER	KEY	23	17

< 4 bytes >

Appraisal

This scheme works, and usefully illustrates much of the behaviour that we want in an abstract data typing mechanism. It also has several major faults and omissions which need to be rectified if it is to be of any serious use. Let us summarize the pros and cons.

PRO:

(i) It is reasonably efficient at run-time. The use of the instance variables REAL and IMAG imposes a run-time overhead of one addition operation compared to ordinary Forth variables.
(ii) It is reasonably efficient in space terms. The memory overhead is only two bytes per object (to store the key), regardless of the size of an object.
(iii) The syntax is simple, transparent and consistent with normal Forth usage, in that operators follow their operands.

CON:

(i) Most serious of all, this implementation does not impose full information hiding. The instance variables REAL and IMAG are present in the private dictionary, and so the addresses of the corresponding data fields can be obtained thus :

```
X REAL . 39256 ok
X IMAG . 39258 ok
```

This would permit a programmer to directly manipulate the data in an object, so violating the principle of abstraction and defeating the purpose of the exercise.

The instance variables need to be hidden even from the private dictionary, and must only be available at the time that the operations are compiled.
(ii) As things stand, objects can only be used in interpretive mode! For example we cannot compile a definition such as

```
: test  X @@ ;
```

The FIND in DO.OPERATION will look for an operation name when **test** is executed, not when it is being compiled. The com-

pilation ends in an error because @@ is not defined as far as Forth is concerned. We need to implement early binding for operations as well as for instance variables.
(iii) There is a run-time overhead caused by the execution of UNLOCK and LOCK. Early binding for operations would eliminate this too.
(iv) The code for @@ and !! is made ugly, contorted and unreadable by the use of the stack to convey the object's base address. This address must be DUPed to provide copies for all the different instance variables, resulting in too many stack contents for comfort, and hence copious and distracting use of ROT, SWAP etc. Do not forget that COMPLEX is a very simple type; when more complicated types are declared this problem will seriously hinder the programming task.
(v) Types defined by this mechanism cannot be nested. It is not possible to declare a new type, one of whose instance variables is a COMPLEX.

In the next sections we shall overcome each of these objections, in some cases by making justifiable trade-offs of efficiency.

Hiding the Instance Variables

The first and major failing, namely the availability of the instance variables, can be overcome easily and at little cost by a slightly more sophisticated approach to the private dictionary structure.

What is required is a second level of "sealing" inside the private dictionary so that the instance variables are locked out when the type declaration is compiled. In order to do this, information about the location of the end of the instance variables must be preserved. This can be accomplished by adding a new syntactic element which separates the instance variables from the operations, at the same time recording the relevant address :

```
TYPE> COMPLEX
2 VAR REAL
2 VAR IMAG
OPS>
: !!  DUP IMAG ROT SWAP !  REAL SWAP ! ;
: @@  DUP REAL @  SWAP IMAG @ ;
ENDTYPE> COMPLEX
```

This new element, which we have chosen to call OPS> to indicate OPerationS, must also create a dictionary link field through which the sealing off of the variables can be performed. In our original diagrammatic notation it would appear :

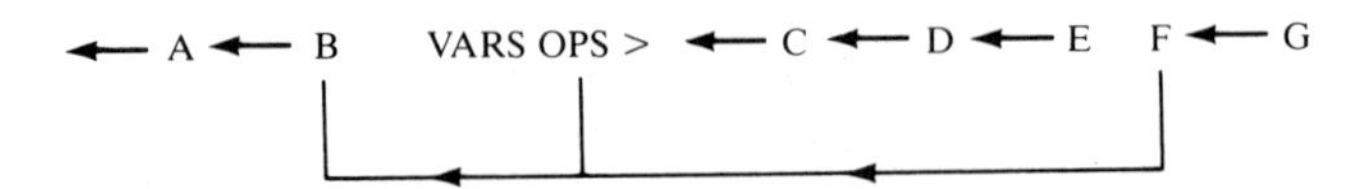

There is however no need for OPS> to compile a full header, and so we shall "fudge" a partial header as follows :

```
VARIABLE OPS

: OPS>     HERE                   -- address following last VAR
           0 C,                   -- make a null name field
           LATEST ,               -- make link field pointing to last VAR
           DUP CONTEXT @ !        -- tell Forth about this dummy header
           N>LINK OPS !   ;       -- preserve dummy's LFA
```

Following the execution of OPS>, the variables are still linked into the dictionary, and so references to them can be correctly compiled into the operations code. But now, using the information preserved in the variable OPS, SEAL can be altered so that the variables are locked out when the type body is sealed :

```
: SEAL  HERE BODY> >LINK PUBLIC @ SWAP !       -- seal private dictionary
        0 LASTPRIVATE @ BODY> >LINK !          -- unlink from main dictionary
        LASTPRIVATE @ BODY> >NAME OPS @ !   ;  -- seal off variables
```

The alteration to the syntax caused by the introduction of OPS> is wholly beneficial, as it forms a clear boundary between the visible and invisible parts of the private dictionary. It is quite permissible to define low level auxiliary operations for use in the operations proper which can be hidden by defining them before OPS> :

```
TYPE> STACK
2   VAR STACKPOINTER
100 VAR STACKBODY
: INC    2 STACKPOINTER +! ;
: DEC   -2 STACKPOINTER +! ;
OPS>
etc............
```

Equally it is possible to reveal some of the instance variables by declaring them after OPS>. If done in a disciplined way this need not violate the principle of information hiding. An example of a "benign" use of visible instance variables would be to provide a field called LINK which allows objects of a type to be "strung" together into linked lists (these will be discussed in full in the next chapter). Rather than putting all the list manipulating code into the type's operations, which would distract from the type's intrinsic properties, it would be better simply to make LINK visible to external list manipulating words.

Shared Variables

Since any arbitrary Forth code may be placed in the body of a type declaration, it must be possible to declare ordinary Forth variables and constants in a type. Let us consider what the effect of this would be. Say that we declared a type

```
TYPE> TEST
2 VAR A
2 VAR B
VARIABLE C
10 CONSTANT D
OPS> ...........
```

The instance variables A and B are "dummies"; they do not represent any storage area, but merely an offset into the storage area of instances of the type to be created later. They represent the local variables inside instances of TEST, of which there may be many copies.

The variable C however is a "real" variable, in that it represents a storage area for a number. Moreover it is not "replicated" whenever a new instance is created. There is only ever one copy of C, sitting in the definition body of TEST. It does however have one interesting property. Being declared inside the encapsulated private dictionary of TEST, it is only visible to objects of type TEST, and not to the rest of Forth. In other words C is a single storage location which is shared by all instances of TEST, but is private to the type.

Such a variable can be used as a "mailbox" by which instances of the same type leave messages for each other. It is a secure mailbox because no other Forth word may alter its value. Since it has been declared before OPS> it cannot even be accessed directly through an instance of TEST, but may only be modified by one of the operations of TEST.

An example of the use of such a construct could be to remember the largest value stored in any instance of a type :

```
TYPE> WIDGETCOUNT
2 VAR WIDGETS
VARIABLE MAXWIDGETS  0 MAXWIDGETS !
OPS>
: W!  DUP WIDGETS !  MAXWIDGETS @ MAX MAXWIDGETS ! ;
..........
```

Objects of type WIDGETCOUNT automatically keep a running maximum value in MAXWIDGETS, which is available to any instance. If later on we added an operation WIDGETGRAPH to produce a bar chart of the widget counts, it would be possible to make the values self-scaling by using the value of MAXWIDGETS, e.g. :

```
: WIDGETGRAPH   WIDGETS @ 100 * MAXWIDGETS @ /  plot etc........
```

Constants can similarly be shared by the instances of a type, as can tables of constants.

Early Binding for Operation Calls

Both the second and third flaws identified above can be eliminated by making the look-up of operations occur at compile-time rather than run-time.

As discovered in the last chapter, it will be necessary to make object/operations "state-smart" so that they behave differently according to whether they are being interpreted or compiled. As before this will involve a slight loss of efficiency during interpretation.

In addition to DO.OPERATION, a word COMPILE.OPERATION is needed, which compiles a reference to the operation rather than executing it. Since both DO.OPERATION and COMPILE.OPERATION need to look into the private dictionary, it will pay to factor out the code which does this :

```
: FIND.OP   BL WORD SWAP                                -- get operation name
            UNLOCK FIND LOCK                            -- find its CFA
            0= IF ." unrecognized operation" ABORT      -- abort if not found
               ENDIF ;                                   ( key --- CFA )

: DO.OP          FIND.OP EXECUTE ;                       ( addr key --- ? )

: COMPILE.OP     FIND.OP                                 ( addr key ---    )
                 SWAP [COMPILE] LITERAL      -- compile obj addr as a literal
                 , ;                         -- and then operation CFA

: DO.OR.COMP     STATE @ IF   COMPILE.OP                 ( addr key ---    )
                         ELSE DO.OP
                         ENDIF ;

: MAKE.INSTANCE    CREATE                    -- make a named instance
                     DUP @ ,                 -- store key into instance
                     2+ @ ALLOT              -- allot storage space
                     IMMEDIATE               -- must execute at compile time
                   DOES> DUP @               -- put PFA and key on stack
                         DO.OR.COMP ;
```

With these changes, it is possible to compile the previous example definition correctly :

```
: test  X @@ ;
```

The CFA of @@ will now be directly compiled into test, and no look-up need be performed when **test** is executed. The address of X is compiled as a literal, and so behaves just as if X were an ordinary variable.

The Object Stack

The first three flaws have been removed at a considerable profit; the efficiency of the implementation has actually been increased by the necessary changes. With the fourth our luck runs out. To clean up the messy stack manipulation code entailed by keeping object addresses on the parameter stack, we shall have to pay back something in the way of efficiency.

The aim is to make the instance variables behave exactly like Forth variables, rather than as offset-adding words which require a base address on the stack. Instead of

```
TYPE> COMPLEX
2 VAR REAL
2 VAR IMAG
OPS>
: !!  DUP IMAG ROT SWAP !  REAL SWAP ! ;
: @@  DUP REAL @  SWAP IMAG @ ;
ENDTYPE> COMPLEX
```

we want to be able to write

```
TYPE> COMPLEX
2 VAR REAL
2 VAR IMAG
OPS>
: !!   IMAG !  REAL ! ;
: @@   REAL @  IMAG @ ;
ENDTYPE> COMPLEX
```

To achieve this, object addresses must be passed somewhere else but on the parameter stack, and whatever method we choose is bound to increase the run-time overhead.

Whether or not this trade-off is justified may be somewhat controversial; "traditional" Forth programmers will probably feel that it is much ado about nothing, and not worth losing any cycles over. Those who take this attitude are unlikely to feel any great need for abstract data types anyway!

The argument for proceeding is this: in addition to compile-time and run-time, one must consider programming and debugging time. The trade-off cuts three ways rather than two. The cleaner code in the second version above will save very significant amounts of programming and debugging time when less trivial types than COMPLEX are being declared. There will also be some *saving* of run-time overhead thanks to the elimination of stack manipulation words such as DUP, ROT and SWAP. Provided that the run-time penalty can be kept small, this is a highly desirable enhancement.

When seeking a method for passing object addresses, several possibilities must be considered.

A very cheap solution would be to put object addresses on the Forth return stack. However a little experimentation should convince us that this is unworkable. Access to the return stack is not sufficiently predictable, particularly when DO...LOOP is used in operations.

The next most obvious solution is to keep the address of the current object in a variable. This would allow VAR to be re-defined thus :

```
VARIABLE OBJ

: OBJ@+    OBJ @ + ;

: VAR   CREATE  SIZE @ ,                              ( n +++  )
                SIZE +!
                IMMEDIATE
        DOES>   @ [COMPILE] LITERAL
                COMPILE OBJ@+  ;
```

Both DO.OP and COMPILE.OP then need to store the object base address into OBJ rather than leaving it on the stack.

Rather than pursue this solution fully though, let us move on to consider the final flaw detected above, namely that these data types cannot be nested. If this flaw is remedied, then the possibility will be raised of objects as variables within other objects. When executing the code for an operation on such an object, an operation upon an embedded object may be encountered. For example,

```
TYPE> TEST
2 VAR   A
COMPLEX B
OPS>
: all@     B @@   A @ ;
ENDTYPE> TEST

TEST X

X all@
```

When B @@ is executed, the address of B will be placed into OBJ and overwrite the address of X, and this will cause a disastrous error when A @ is performed. Clearly the address of X must be preserved and restored after the operation on B has finished. So the correct solution, if we are to anticipate a later enhancement to allow nesting of types, is to use a dedicated *stack* rather than a variable to pass object addresses.

Having eliminated both the parameter and return stacks as candidates, the only option is to create a third stack. This stack need not be large as it only has to accommodate as many addresses as the maximum level of nesting to be permitted ; 20 levels would be quite lavish.

This object stack, as we shall call it, does not need all of the various operators that are theoretically possible. These three will suffice :

```
OPUSH   -- push top of parameter stack to object stack.
OPOP    -- pop the object stack, losing the top item.
OCOP+   -- copy top of object stack to parameter stack and add.
```

The object stack can be created in high-level Forth as follows, though it is highly desirable to implement the operators in machine code to minimize the overhead.

```
40 CONSTANT MAXNEST

CREATE OSTACK  HERE MAXNEST +  ,   MAXNEST ALLOT
```

This creates a stack descending from high memory, with the stack pointer stored in OSTACK. The operators are

```
: OPUSH  OSTACK -2 OVER +!  @ !  ;
: OPOP   2 OSTACK  +!  ;
: OCOP+  OSTACK @ @ +  ;
```

No test for empty stack is required as the way we shall use the operators precludes stack underflow. A test for stack overflow could be included but will be omitted here for simplicity's sake. These operators are used solely by the implementation, and must not be used explicitly by the programmer.

An operation will now be performed with the address of its object on the top of the object stack, rather than the Forth parameter stack. Both DO.OP and COMPILE.OP need altering to do the necessary push and pop :

```
: DO.OP        FIND.OP SWAP  OPUSH EXECUTE OPOP ;        ( addr key --- ? )

: COMPILE.CALL  COMPILE OPUSH , COMPILE OPOP ;           ( CFA ---   )

: COMPILE.OP    FIND.OP SWAP [COMPILE] LITERAL           ( addr key ---    )
                COMPILE.CALL ;
```

VAR can be re-defined as :

```
: VAR    CREATE  SIZE @ 2+ ,                             ( n +++  )
                 SIZE +!
                 IMMEDIATE
         DOES>   @ [COMPILE] LITERAL
                 COMPILE OCOP+  ;
```

The cost in efficiency can now be quantified. For a compiled operation, there is a run-time penalty of one OPOP and one OPUSH per call, plus an additional penalty per instance variable used in the operation. This penalty is the time difference between the simple + used previously and OCOP+ (roughly the time for three fetches in our high-level implementation). If the object stack operators are carefully coded in assembler this can be made quite small.

An important consideration is stack security. A stack requires some mechanism to ensure against underflow (popping from an empty stack) and overflow (pushing to a full stack).

The above scheme eliminates the problem of stack underflow because OPOP can never be executed unless preceded by a corresponding OPUSH. The converse is not true however; OPUSH can be executed without a following OPOP in the special case where an operation ends in an error which causes an ABORT. In such a case one extra item is left on the object stack, and repeating the process will eventually lead to overflow. The proper way to surmount this problem is not through a time-consuming test for overflow, but rather to modify ABORT so that it resets the object stack pointer, as it already does for the parameter and return stacks. How this is to be done will vary between Forth implementations; most good ones will have provided vectored execution for ABORT to allow easy replacement by a user definition :

```
: ORESET     OSTACK DUP MAXNEST + SWAP ! ;

: NEWABORT   ORESET ABORT ;

' NEWABORT   UABORT !         -- vector for user ABORT
```

Something along these lines could also be used in Forth-79 versions to trap the previously identified fatal error condition caused by an object name without a following operation, by making ABORT do a LOCK :

```
: NEWABORT   LOCK ORESET ABORT ;
```

Here we are opening a dangerous can of worms though. This code will get executed when *any* Forth error, not merely a type related error, occurs. It is imperative that STASH always contains a valid vocabulary address, so it will need to be initialized at boot time. Moreover, this solution may interfere with the way FORGET and other vocabulary manipulation words like PROTECT or FREEZE work. Experiment with it if you will, but the solution presented later, using type DEBUG>, is much safer.

Nesting Types

The full power of abstract data types can only be realized if types may be nested; that is, if a previously defined type can be used as a component of a new type. This ability greatly helps with program design by encouraging the decomposition of problems into data/operation modules which can be reused. It enhances Forth's already considerable ability to produce very compact programs by reusing code.

By opting, at some expense, for the use of a stack to pass object addresses to operations, we have laid the ground for such nesting. What remains to be done is to alter the behaviour of type defining words so that they can create either instances or instance variables according to context.

To provide this dual behaviour, it is necessary to extend the concept of "state" in Forth. Currently Forth is a four state system; it is either compiling or interpreting (signalled by the value of STATE) from disk or terminal (signalled by BLK). An extra pair of states is required, namely "in" or "not-in" a type declaration. By testing this state, a type defining word can perform the correct action.

A variable called IN.TYPE.DEF? will be used as a flag to distinguish the states. TYPE> sets IN.TYPE.DEF? to TRUE and ENDTYPE> sets it to FALSE.

When a previously defined type is used to create an instance variable rather than an instance, a very different behaviour is required. No space needs to be allocated. Instead the offset and type key must be recorded in the variable name. When such a variable name is executed it must, as usual, add its offset to the base address of the current object. However the resulting address is not a data field but the address of an embedded object. This is then itself treated as the current object, with its own operation to be applied.

A word MAKE.INSTVAR is needed, which behaves like a hybrid of VAR and MAKE.INSTANCE. It requires, like MAKE.INSTANCE, the PFA of a type defining word to be on the stack :

```
: MAKE.INSTVAR  DUP 2+ @              -- get storage size ( PFA +++  )
                SWAP @                -- get key
                CREATE  ,             -- store key
                        SIZE @ ,      -- store offset (no key)
                        SIZE +!       -- bump size
                        IMMEDIATE
                DOES>   DUP @              -- get key
                        SWAP 2+ @          -- get offset
                        [COMPILE] LITERAL  -- code to add offset...
                        COMPILE OCOP+      -- ...at run-time
                        FIND.OP            -- ...and treat result
                        COMPILE.CALL ;     -- ...as an object
```

All that remains is to modify TYPE> and ENDTYPE> :

```
VARIABLE IN.TYPE.DEF?

: TYPE>      LATEST PUBLIC !          -- NFA of first public word
             CREATE                   -- make header
             HERE LASTPRIVATE !       -- PFA of this new word
             0 SIZE !
             TRUE IN.TYPE.DEF? ! ;

: ENDTYPE>   LATEST                   -- NFA of first private word
             CREATE SEAL              -- make typeword and seal
               , SIZE @ ,             -- store key and storage size
             FALSE IN.TYPE.DEF? !
             DOES> IN.TYPE.DEF? @
                   IF   MAKE.INSTVAR
                   ELSE MAKE.INSTANCE
                   ENDIF ;
```

It may be helpful to examine an example, using the following type declaration, to see how this works :

```
TYPE> TEST
2 VAR A
COMPLEX B
OPS>
: com!   B !! ;
: scal!  A ! ;
: all@   B @@   A @ ;
ENDTYPE> TEST

TEST X

23 45 X com!
99 X scal!
```

When **com!** is executed, the address of X is initially on the object stack. B adds its offset to the address of X, and then pushes this address on top of X and executes !! as if B were the current object.
The internal structure of the instance X looks like this :

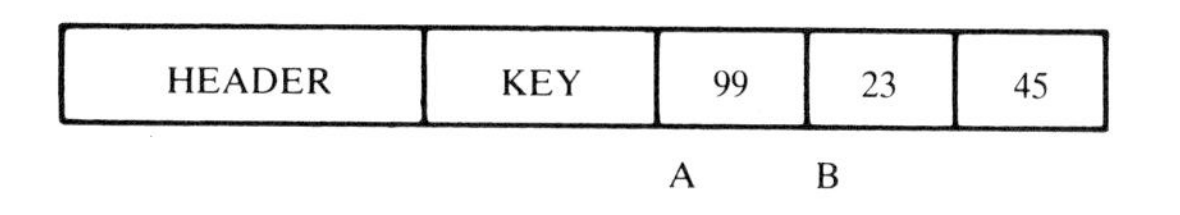

Notice that the "complex" part of X has no key field. In an early binding implementation it is not necessary that the embedded "complex" object referred to by B should be a fully fledged object, complete with a key field.

The key is stored instead in the variable name B, so the type is always known at compile time. This saves a lot of memory; there is still only a two-byte overhead no matter how complicated the structure of the object.

On the other hand the run-time speed overhead is greater than that for a simple object. In addition to the OCOP+ associated with any instance variable, there is the overhead of an OPUSH and OPOP compiled by COMPILE.CALL, which puts the calculated address of the embedded object onto the object stack.

Summary

After so many fundamental changes it would be useful to put all the code together for reference. Also a little re-factoring can be done to smarten up the code :

```
-- Working variables for object compiler
VARIABLE SIZE          -- Holds storage size of type
VARIABLE OPS           -- Holds address of end of ops vocabulary
VARIABLE STASH         -- Temporary store for current vocabulary
VARIABLE PUBLIC        -- Holds link to ordinary dictionary
VARIABLE LASTPRIVATE   -- Holds address of last word in type
VARIABLE IN.TYPE.DEF?  -- Flag; are we in a type definition?

-- Make a third stack to hold current object's address ; its size
-- determines how deeply type definitions may be compounded

40 CONSTANT MAXNEST

CREATE OSTACK  HERE MAXNEST + ,   MAXNEST ALLOT

-- Push parameter stack to object stack
: OPUSH   OSTACK -2 OVER +!  @ !  ;                          ( n --- )

-- Pop object stack and discard
: OPOP    2 OSTACK  +!  ;                                    ( --- )

-- Copy top of object stack and ADD to top of parameter stack
: OCOP+   OSTACK @ @ +  ;                                    ( n --- n )

-- Compile offset into instance variable name, then bump the total
: OFFSET   SIZE @ 2+ , SIZE +! ;                             ( size --- )

-- Purely for brevity
: COMPLIT   [COMPILE] LITERAL ;

-- Compile code to add offset into object body
: COMPILE.ADDOFF   COMPLIT COMPILE OCOP+ ;

-- Create a new instance variable of 'size' bytes
: VAR    CREATE  OFFSET                                      ( size ---  )
                 IMMEDIATE
         DOES>   @ COMPILE.ADDOFF ;

-- Open a type declaration
: TYPE>   LATEST PUBLIC !              -- NFA of last public word
          CREATE                       -- Make a header
          HERE LASTPRIVATE !           -- Store its PFA
          0 SIZE !                     -- Initializations
          TRUE  IN.TYPE.DEF? ! ;

-- Mark boundary which hides the instance variables
: OPS>    HERE                  -- Address following last VAR
          0 C,                  -- Make dummy name field
          LATEST ,              -- Link field points to last VAR
          DUP CONTEXT @ !       -- Let Forth know about dummy word
          N>LINK OPS ! ;        -- Save its LFA
```

```
-- Save current vocabulary; set operations vocabulary
: UNLOCK   CONTEXT @ DUP  @ STASH !  ! ;                 ( key --- )

-- Restore current vocabulary
: LOCK     STASH @ CONTEXT @ ! ;

-- Look up an operation in its type vocabulary          ( key --- CFA )
: FIND.OP  BL WORD SWAP                                -- Get operation name
           UNLOCK FIND LOCK                            -- Find it
           0= IF ." unrecognized operation" ABORT      -- abort if not found
              ENDIF ;

-- Execute an operation if found
: DO.OP    FIND.OP SWAP OPUSH EXECUTE OPOP ;            ( addr key --- ? )

-- Compile operation calling sequence
: COMPILE.CALL  COMPILE OPUSH , COMPILE OPOP ;          ( CFA ---  )

-- Look-up operation and compile it
: COMPILE.OP    FIND.OP SWAP COMPLIT                    ( addr key ---  )
                COMPILE.CALL ;

-- Fetch size field contents from instance variable or type
: SZ@  2+ @  ;                                          ( addr --- size )

-- Create an instance variable of a predefined type   ( addr ---  )
: MAKE.INSTVAR    DUP SZ@                          -- Get size
                  SWAP @                           -- Get key
                  CREATE , OFFSET                  -- Store key and size
                         IMMEDIATE
                  DOES> DUP @ SWAP SZ@ 2-           -- Get key and offset
                        COMPILE.ADDOFF             -- Compile code....
                        FIND.OP                    -- to treat as.....
                        COMPILE.CALL ;             -- an object.

-- Compile or interpret an operation according to state
: DO.OR.COMP    STATE @ IF   COMPILE.OP                 ( addr key ---  )
                       ELSE DO.OP
                       ENDIF ;

-- Create a new instance of a type                      ( addr ---  )
: MAKE.INSTANCE   CREATE DUP @  ,          -- Store key into instance
                         SZ@ ALLOT         -- Allot its storage
                         IMMEDIATE
                  DOES> DUP @ DO.OR.COMP ;

-- Juggle dictionary pointers to seal the type body
: SEAL    HERE BODY> >LINK  PUBLIC @ SWAP !                 (  ---  )
          LASTPRIVATE @ BODY> >NAME OPS @ !
          0 LASTPRIVATE @ BODY> >LINK  ! ;

-- Close type declaration
: ENDTYPE> LATEST CREATE SEAL               -- Close the body
                   , SIZE @ ,               -- Store key and size
                   FALSE IN.TYPE.DEF? !
              DOES>  IN.TYPE.DEF? @ IF   MAKE.INSTVAR
                                   ELSE MAKE.INSTANCE
                                   ENDIF ;
```

Only the words TYPE>, OPS>, ENDTYPE> and VAR are to be used in user programs; all the others are for internal use by the compiler and will result in disaster if applied casually. For this reason the internal words would best be locked away in a conventional Forth vocabulary (perhaps called TYPEVOC) for a polished presentation. This will not be done here, as the purpose of this book is tutorial, and it would obscure the meaning of the code too much.

Using Abstract Data Types

The system we have arrived at overcomes all the previously raised objections, and provides an effective and reasonably efficient system of abstract data types. Before considering whether any further enhancements are desirable, it would be useful to spend a while considering the applications for abstract data types, on the assumption that this is the best way to spot significant deficiencies.

Abstract data types could be used to extend the range of simple data types provided by Forth. The private dictionary mechanism permits the "overloading" of operators, that is the use of the same operator for different but analogous operations. For example, many languages which support floating point arithmetic overload the basic arithmetic operators +,-,* and / so that they work on integers or floating point numbers. There is as yet no standard for Forth floating point, but in any case the structure of Forth demands that different operators (e.g. F+,F-,F* and F/) be used to perform FP arithmetic.

By implementing a floating point package as a type declaration, it would be possible to create a type FLOAT which uses the normal Forth operators !, @, +! to manipulate floating point variables. However such a system suffers from severe drawbacks.

Firstly our typing mechanism only produces typed variables, and cannot produce typed literals. For example a COMPLEX literal is still just two numbers on the stack, e.g. 23 45. Even if we overload the floating point variable operators !, @ etc. we must still use F+, F- etc. on literal expressions. This produces a confusion which is wholly counter productive.

Secondly, the operation look-up mechanism effectively restricts us to unary operations. Since typed variables are active objects which demand an operation name to follow them in the input stream, it is very difficult (and extremely messy) to devise a method of applying a binary operation to two variables.

The truth is that Forth is still too low-level (and most of us prefer it that way) to handle gracefully expression oriented arithmetic on structured data types. The only effective solution would be to produce a sophisticated outer interpreter which parses input expressions more fully than Forth does. Such an interpreter would most naturally use infix rather than Reverse Polish notation. The result would be to turn Forth into a more conventional language resembling Basic, or better, Pascal. While there are eloquent proponents of this solution (e.g. the authors of Magic-L) it is far too involved, not to mention controversial, for a book such as this about Forth.

As a result of these deliberations we must conclude that extending the range of simple data types is not a very good use for abstract data types in Forth.

The classic application of abstract data types is for implementing complex data structures which are intrinsically fragile. Many of the data structures routinely used in system progamming, such as stacks, queues, buffers, tasks, resource managers and monitors, are very easily disrupted. The corruption of a single pointer value often results in the loss of the whole data structure and total system failure.

Abstract data types are the ideal tool for implementing such structures. The restriction of access which an abstract data type entails is the best guarantee of integrity, and is especially beneficial in Forth, which otherwise allows such unrestricted access to memory. Moreover, clever use of abstract data types can lead to reusable code. A library of types can be built up so that whenever, say, a queue is required, it can be produced from an existing template.

Returning to the first example in this chapter, let us see one possible way to create a type STACK.

```
TYPE> STACK
  2       VAR STACKPTR
  50 2*  VAR STACKBODY
  : INC    2 STACKPTR +! ;
  : DEC   -2 STACKPTR +! ;
OPS>
  : INIT       STACKBODY STACKPTR ! ;
  : EMPTY?     STACKPTR @ STACKBODY = ;                      (   --- flag)
  : PUSH       STACKPTR @ ! INC ;                            ( n ---     )
  : POP        EMPTY? NOT IF    DEC STACKPTR @ @             (     --- n )
                          ELSE ." stack empty!"
                          ENDIF ;
ENDTYPE> STACK
```

This version of STACK has a test for empty stack, and will not allow POP on an empty stack. Of course a similar test for stack full could be easily added. Stacks of this type all have the same size, namely 50 single integers deep. Any number of new stacks can be created by merely saying

```
STACK A   STACK B   STACK C

A INIT   B INIT C INIT

33 A PUSH           etc.....
```

INC and DEC have been hidden, by defining them before OPS>, so the user cannot directly manipulate the stackpointer.

Note that each new stack has to be explicitly initialized before use, and the penalty for omitting this step is disaster! This immediately suggests a significant enhancement; instances could be automatically initialized when they are created, by executing a special operation which is always called INIT.

Objects of type STACK can be used as instance variables in defining new types. For instance in a multitasking system, task descriptors could be represented by a type which contains one or more stacks.

On examining the declaration of STACK, we can see two areas in which the mechanism could be made much more powerful.

Firstly, type STACK is limited to stacks of 50 integers, though none of the operations depend upon the size of the stack (FULL? would, if implemented). It would be very handy if we could create a single type STACK which could create stacks of any size; a skeleton definition of the archetypical stack.

Secondly, type STACK is limited to holding single integer items. If we wanted to create a stack of some predefined type of item, say COMPLEX, we would need to start again. Moreover we lack any mechanism for creating an array of structured items such as COMPLEX, which would be required for the STACKBODY.

In the next sections we shall examine these possible enhancements in more detail.

Initialization

The creation of automatically initialized instances is not difficult. A simple, though rather inflexible, way is to replace ALLOT wherever it occurs with a word which initializes the alloted space to zeroes. This word will do the trick:

```
: ALLOTZ    DUP HERE SWAP 0 FILL ALLOT ;             ( size --- )
```

In a large number of cases this is all the initialization that is required. There will be occasions however where more complex initializations are required, and in such cases it would be preferable to force the programmer to explicitly include a word among the operations which performs the desired actions. This encourages the good programming practice of always considering the initial conditions of a program.

The word MAKE.INSTANCE can be altered so that in addition to allotting space for a newly created instance, it also *executes* the operation INIT on that instance. The changes required are :

```
: INITIALIZE  SWAP OPUSH                           ( addr key ---  )
              UNLOCK  " INIT"  FIND LOCK        -- find op called INIT
              IF   EXECUTE
              ELSE DROP
              ENDIF  OPOP ;

: MAKE.INSTANCE  CREATE HERE SWAP                 ( addr ---  )
                        DUP @ DUP ,               -- store key into instance
                        SWAP SZ@ ALLOTZ           -- allot its storage
                        INITIALIZE                -- perform initialization
                        IMMEDIATE
                 DOES> DUP @ DO.OR.COMP ;
```

Note that the definition of INITIALIZE requires the use of a word which is not in either the 79 or 83 standards, namely " which is used to create a string literal in a colon definition. Most good Forth implementations will have such a word (which might be called something else, such as LIT"). Without it this code cannot work, and so here is a high-level definition of " which you can use if it is not supplied with your system :

```
: (")  R@ DUP C@ 1+ R> + >R ;          -- runtime code pushes addr and skips
                                                  ( --- addr )

: "    34 WORD                        -- get string up to next " char
       DUP C@ 1+                      -- extract the count
       >R                             -- save count
       HERE 2+ R@ CMOVE>              -- move string to avoid corruption
       COMPILE (")                    -- compile runtime code
       R> ALLOT ;   IMMEDIATE         -- adjust HERE past the string
                                                  (  +++  )
```

Note that this version of " is not state-smart and must only be used inside colon definitions. Note also that ", like WORD, returns a single address (as opposed to an address and count) which is what FIND wants.

In any case, the initialization mechanism cannot be made to work in Forth-79, because the 79-FIND cannot take a string argument. 79-FIND requires a word in the input stream at run-time, so the best we could manage would be :

```
: MAKE.INSTANCE   CREATE HERE SWAP         ( addr ---    )
                    DUP @ DUP ,          -- store key into instance
                    SWAP SZ@ ALLOTZ      -- allot its storage
                    DO.OP                -- look for an operation
                    IMMEDIATE
                  DOES> DUP @
                        DO.OR.COMP   ;
```

and then create initialized instances as follows :

```
COMPLEX X INIT        or        STACK B INIT
```

This not only spoils the syntax but leaves open the possibility of accidentally omitting the INIT; it is no great improvement on explicitly calling INIT.

In general it is best to make INIT the last operation to be defined. The reason is simply that this allows all the previously defined operations to be used in the implementation of INIT; some of them will typically be data store operations which can be used to initialize data values.

Taking COMPLEX as an example, we could initialize as follows :

```
TYPE> COMPLEX
  2 VAR REAL
  2 VAR IMAG
OPS>
  : INIT  0 REAL ! 0 IMAG ! ;
  : com!  IMAG ! REAL ! ;                     ( n n ---    )
  : com@  REAL @ IMAG @ ;                     (    --- n n )
ENDTYPE> COMPLEX
```

But by placing INIT at the end, we can use **com!** to define it :

```
TYPE> COMPLEX
  2 VAR REAL
  2 VAR IMAG
OPS>
  : com!  IMAG ! REAL ! ;
  : com@  REAL @ IMAG @ ;
  : INIT  0 0 com! ;
ENDTYPE> COMPLEX
```

A point worthy of comment is the curious body of INIT in this case. It may not at first be obvious what is happening when **0 0 com!** is executed, as there appears to be no object for **com!** to work on. The answer is of course that the address of the current object is always on the object stack when INIT (or any other operation) is executed; the object of **com!** is *implied* to be that pointed to by the object stack top. This can be a source of puzzlement when reading operation code which uses prior operations from the same type. Taking a lead from Smalltalk-80, we could invent a pseudo-object called SELF to represent the current implied object. The definition of SELF is not too taxing :

```
: SELF ;
```

In other words SELF is completely redundant, given our scheme of object reference, but it may be included for readability :

```
: INIT   0 0 SELF com! ;
```

These examples also raise an interesting point about operation order. A strong argument against the casual use of overloaded operators is that they deny access to the original operators for all the operations defined after them. If for example one chose to call **com!** simply **!** (thus overloading **!**), then any other operations which need the standard Forth **!** must be defined earlier, before the overloading occurs. This constraint can become vexing if multiple overloaded operators are declared.

Though it may seem like a case of belt and braces, it is probably a good idea to include both the initialization mechanisms discussed here, i.e. to replace ALLOT with ALLOTZ as well as using INIT. The effect on efficiency is negligible as instance creation time is not usually critical, and the benefit is that everything gets initialized to zero by default, unless a specific INIT adds further actions.

Inheritance

The variable size requirement is not so easy to satisfy. It is easy enough to modify MAKE.INSTANCE so that it takes a parameter from the stack which determines the actual size of object to be created. However, this introduces several highly undesirable side effects. Objects now need to have an embedded size field for each such variable sized field, to show what size they were actually instantiated to. This in turn horribly complicates the nesting of such types, as well as imposing a memory overhead.

On balance the extra complication is not justified, because a very simple if slightly less powerful alternative exists. Instead of making a single variable sized type, we could allow types to *inherit* the operations of a parent type. A single generic type GENSTACK would contain all the stack operations which do not depend upon size. Then differently sized types like STACK20 and STACK80 could be defined, which inherit nearly all their operations from GENSTACK, and hence have very simple declarations. This feature can be had for absolutely no cost, because of the way our private dictionary works. Instead of putting a zero in the last link field, to act as a "stopper", we can put the key of the parent type. The private dictionary of the parent type then becomes appended to that of its child, forming a chain.

Only a single new word is required, called INCLUDE>. INCLUDE> can be placed anywhere inside a type declaration, and must be followed by the name of the parent type; only one such type may be included. Here is the definition of INCLUDE> :

```
VARIABLE INHERIT

: INCLUDE>   ' >BODY @ INHERIT ! ;          -- store key of parent class
                                            -- in INHERIT
```

SEAL needs to be modified to use the contents of INHERIT instead of zero:

```
: SEAL    HERE BODY> >LINK
          PUBLIC @ SWAP !                    -- seal private dictionary
          INHERIT @ LASTPRIVATE @
          BODY> >LINK !                      -- link to parent dictionary
          LASTPRIVATE @ >NAME OPS @ !   ;    -- seal off variables
```

In addition, INHERIT needs to be initialized to zero inside TYPE>, so that the use of INCLUDE> remains optional; in its absence the system behaves as before. Using these definitions we can produce various sized types from a single parent type :

```
TYPE> GENSTACK
  2 VAR STACKPTR
  0 VAR STACKBODY
  : INC    2 STACKPTR +! ;
  : DEC   -2 STACKPTR +! ;
OPS>
  : INIT      STACKBODY STACKPTR ! ;
  : EMPTY?    STACKPTR @ STACKBODY = ;                  ( --- flag)
  : PUSH      STACKPTR @ ! INC ;                        ( n ---    )
  : POP       EMPTY? NOT IF   DEC STACKPTR @ @          (   --- n )
                         ELSE ." stack empty!"
                         ENDIF ;
ENDTYPE> GENSTACK

TYPE> STACK10
  2 VAR STACKPTR
  10 2*  VAR STACKBODY
OPS>
INCLUDE> GENSTACK
  : FULL?     STACKPTR @ STACKBODY - 18 > ;             ( --- flag)
ENDTYPE> STACK10
```

This works because the instance variable names in reality represent offsets and not absolute addresses. By declaring the same sized variables in the same order in STACK10, the names used in GENSTACK operations correctly refer to the instance variables in STACK10. This means that only one field, the last one *declared*, may be variably sized in this way. Note that the STACKBODY in GENSTACK is declared as a dummy variable of length 0; this is because only its offset matters, as we are never going to create any instances of GENSTACK, only of its children. This could be made even plainer by declaring DUMMY as a constant, value 0, and saying :

```
DUMMY VAR STACKBODY
```

STACK10 declares a new operation, FULL?, which does depend upon size, and so must be declared anew in each child type. However there are several limitations on the declaration of operations in the child type. STACK10 cannot refer directly at compile-time to any of the operations in GENSTACK, because the latter's private dictionary is locked up. It is also impossible for operations in GENTYPE to refer forward to the properties of its children's instances; that is, it is not possible to declare FULL? in GENSTACK and somehow feed it the child's size at run-time, because early binding requires this knowledge too soon.

The first limitation can be eased by a different way of declaring STACK10:

```
TYPE> STACK10
  GENSTACK  STK
  10 2* VAR STACKBODY
OPS> INCLUDE> GENSTACK
  etc....
```

In this case we are using an instance variable of type GENSTACK to provide the body of STACK10 (it will be slightly less efficient as a result). The variable STK will accordingly have all the operations of GENSTACK available to it. The variable STACKBODY now overlays or "aliases" the dummy variable of the same name in GENSTACK; since it starts at the same relative offset it becomes an alias for the original. This form enables us to put much of the code for FULL? back into GENSTACK :

```
TYPE> GENSTACK
  2 VAR STACKPTR
  DUMMY VAR STACKBODY
  : INC    2 STACKPTR +! ;
  : DEC   -2 STACKPTR +! ;
OPS>
  : INIT     STACKBODY STACKPTR ! ;
  : EMPTY?   STACKPTR @ STACKBODY = ;                    ( --- flag)
  : PUSH     STACKPTR @ ! INC ;                          ( n ---   )
  : POP      EMPTY? NOT IF   DEC STACKPTR @ @            (   --- n )
                       ELSE ." stack empty!"
                       ENDIF ;
  : ITEMS    STACKPTR @ STACKBODY - 2 /  ;               (   --- n )
ENDTYPE> GENSTACK

TYPE> STACK10
  GENSTACK  STK
  10 2*  VAR STACKBODY
OPS> INCLUDE> GENSTACK
  : FULL?   STK ITEMS 9 > ;                              ( --- flag)
  : PUSH    FULL? NOT IF    STK PUSH                     ( n ---   )
                      ELSE ." stack full "
                      ENDIF ;
ENDTYPE> STACK10
```

The redefinition of PUSH illustrates another benefit of this technique; any operation can be "overidden" by a new version defined in terms of the old.

This is about as far as we can go without introducing a great deal more complexity. It is perfectly possible to introduce full sub-class inheritance of the kind seen in Smalltalk-80, which involves the inheritance of the instance variables as well as operations from a parent or super-class. However, it requires a far more complex private dictionary structure which can be opened and closed at compile-time, and will not be dealt with here. Interested readers can find an implementation in (Pountain, 1986).

As a footnote, INCLUDE> can be used to import debugging routines during the development of programs, which are then left out of the finished code. Two useful candidate routines are one to list the contents of the private dictionary, and one to unscramble the dictionary for Forth-79 users whose FIND is not fail safe (see earlier references). Here is a sample DEBUG to which many other operations could be added to taste :

```
TYPE> DEBUG
OPS>
  : OPLIST   OSTACK @ @ @ UNLOCK VLIST LOCK ;         -- list operations
  : DAMN!    LOCK ;                       -- lock an open private dictionary
ENDTYPE> DEBUG
```

(This also demonstrates that it is legal to make a type with no data storage, though it would make no sense to create instances of it.) Though only one INCLUDE> can be used per type, any number of types can include DEBUG, and they can be chained. For example if GENSTACK includes DEBUG, then STACK10 and any other offspring inherit it too.

Array-of

The last of our suggested enhancements is at once the hardest, and also the most worthwhile. The facility to create arrays of objects, both as instances and as instance variables would open up whole new areas of application.

While the private dictionary scheme has considerable charm in terms of simplicity, elegance and compatibility with normal Forth, it also has a major weakness. At the moment we have no way to create headerless, anonymous objects. In some applications (e.g. graphics, discrete simulations) we may wish to create huge numbers of identically structured objects, and the memory overhead of a full Forth header (at least six bytes) per object will be quite unacceptable. Also it may be impossible or undesirable to generate unique names for all these objects (Forth is in any case hopeless for creating such names, because CREATE cannot take a string argument. A future standard should consider modifying CREATE in the same way that FIND was modified in Forth-83.)

It is possible to modify the current scheme to separate headers from object bodies, so that the header becomes a normal Forth variable containing a pointer to the body. This would merely involve removing the CREATE from MAKE.INSTANCE, and having it return the address of the new instance on the stack. However this only displaces the problem.

Structured arrays solve this problem very well. If there is a need for 1000 objects of type BITMAP, merely create an array of 1000 elements, which has only one header and one name. The individual BITMAPs are accessed by indexing. By declaring suitably large arrays, we can even simulate the dynamic allocation and deallocation of objects, using linked lists rather than moving data.

After some experimentation it emerges that the current version of TYPE> is not strong enough to accommodate the creation of arrays of objects within its syntax. A new constructor word ARRAY-OF is needed, which is "superstate-smart" like a type defining word, so that we can say either

```
20 ARRAY-OF COMPLEX X
```

or

```
TYPE> EVENTQUEUE
  2 VAR HEAD
  2 VAR LENGTH
  100 ARRAY-OF EVENT QBODY
OPS> etc.......
```

ARRAY-OF will be, like ENDTYPE>, a second order defining word. It will, also like ENDTYPE>, have two main components to cater for the two states of IN.TYPEDEF? The code compiled by these two components will however be rather more complicated than that compiled by MAKE. INSTANCE and MAKE.INSTVAR, because the indexing calculations have to be performed in order to find out which element of the array is to become the current object.

The question of binding becomes more complex too. When indexing an array, three different binding times are possible. In "full" early binding, the index and the array address are both known at compile-time, and so the actual address of the indexed element can be computed and compiled. In "half" early binding, the array address is known at compile-time but the value of the index is not. In this case we must compile the array address and code to add the index at run-time. In late binding, which is needed to enable arrays to be used in interpreted mode, neither array address nor index are known until run-time, and all calculations are performed then.

There is not enough space to develop ARRAY-OF in the same incremental fashion that we have being following so far. The code is presented as a finished whole, with some explanation of its use. The code is not easy to understand (nor was it easy to write!), particularly in those sections which do the compiling. Anyone who has scrupulously followed the story so far may consider this the end of term test!

The code cannot be understood without knowledge of the data structures involved, which are depicted at the head of the source code. The width of an array simply means the size of an element. Length means the number of elements.

ARRAY INSTANCE

	PFA			
HEADER	KEY	LENGTH	WIDTH	ELEMENTS . . .
	0	2	4	6 byte offset from PFA

ARRAY. VAR NAME

	PFA			
HEADER	KEY	LENGTH	WIDTH	OFFSET
	0	2	4	6 byte offset from PFA

Here is the code :

```
-- Calculate element address.                        ( index pfa +4 width ---        )

  : INDEX+     ROT * + ;

-- Interpret operation on array element.             ( index pfa key ---     )

  : ARRAY.DO.OP     FIND.OP                         -- get operation CFA.
                    ROT ROT
                    4 + DUP @                       -- get width of array.
                    INDEX+                          -- calculate element address.
                    OPUSH EXECUTE OPOP ;            -- do it.

-- Permit full early binding for an index known at compile time.
-- Used as in  VAL[ 3 ] BIGARRAY op

      VARIABLE VAL                                  -- flag full early binding

      : VAL[    TRUE  VAL !  [COMPILE] [  ; IMMEDIATE -- set flag; stop compiling

      : VAL    FALSE VAL !   ;                      -- reset flag

-- Compile operation on array element.               ( <index> pfa key ---     )
-- The index may be present at compile time as a value on the stack (VAL = TRUE)
-- or not (VAL = FALSE).

  : ARRAY.COMP.OP   FIND.OP >R                      -- get op CFA and stash.
                    4 + DUP @                       -- get width.
                    VAL @ IF    INDEX+ COMPLIT      -- compile element address
                          ELSE SWAP COMPLIT COMPLIT -- compile width and pfa
                               COMPILE INDEX+       -- and code to index later.
                          ENDIF
                    R> COMPILE.CALL                 -- compile op call.
                    ~VAL  ;                         -- reset VAL.
```

```
-- Do or compile array op according to STATE          ( index pfa key ---     )

  : ARRAY.DO.OR.COMP    STATE @   IF    ARRAY.COMP.OP
                                  ELSE ARRAY.DO.OP
                                  ENDIF ;

-- Create an instance variable which is a typed array.     ( len pfa ---      )

  : ARRAY.VAR    CREATE                              -- make header.
                   DUP @ , OVER ,                    -- store key and length.
                   SZ@ DUP ,                         -- store width.
                   *                                 -- total size = len * width.
                   OFFSET                            -- store offset & bump total
                   IMMEDIATE
                 DOES>
                   DUP @                             -- get key.
                   FIND.OP >R                        -- get op CFA and stash it.
                   DUP  6 + @ 2-                     -- get offset.
                   SWAP 4 + @                        -- get width.
                   VAL @ IF    INDEX+ COMPILE.ADDOFF -- compile indexing code.
                         ELSE SWAP COMPLIT COMPLIT   -- comp width and pfa...
                              COMPILE INDEX+         -- code to add index...
                              COMPILE OCOP+          -- and code to offset it.
                         ENDIF
                   R> COMPILE.CALL¯VAL   ;


-- Make an array object.                                   ( len pfa ---      )

  : MAKE.ARRAY    CREATE                             -- make header.
                    2DUP @ , ,                       -- store key and length.
                    SZ@ DUP ,                        -- store width.
                    SWAP *  ALLOTZ                   -- allot space.
                    IMMEDIATE
                  DOES>
                    DUP @ ARRAY.DO.OR.COMP   ;


-- Create an array object or variable.                          ( len ---    )

  : ARRAY-OF     ' >BODY                             -- get type pfa
               IN.TYPEDEF? @  IF    ARRAY.VAR
                              ELSE MAKE.ARRAY
                              ENDIF  ;

-- End.
```

You will notice that, for the sake of clarity and speed, no bounds checking has been incorporated into these definitions. The purpose of storing the length in an array object (which is not used in the above code) is to allow such code to be added if desired. The place to perform such checks would be in INDEX+, for example,

```
  : INDEX+     ROT ROT 2DUP                   ( index pfa +4 width ---          )
                OVER 0< >R                      -- negative index?
                2- @ <                           -- length greater than index?
                R> NOT AND IF    ROT ROT * +
                           ELSE ." Array index out of bounds " ABORT
                           ENDIF ;
```

ARRAY-OF is used as follows. A new array can be created by, say,

```
10 ARRAY-OF COMPLEX FRED
```

the length of the array being provided as a parameter on the stack. The elements of FRED behave just like objects of type COMPLEX; ARRAY-OF does not create any new operations. Array elements are accessed by supplying an index value on the stack, before the array name (it is not permissible to use the array name on its own for any purpose). The indices start at 0, so 3 FRED is the fourth element of FRED.

The element 3 FRED responds to any of the operations defined for COMPLEX, and is in effect an anonymous instance of COMPLEX. To store into element 3 of FRED :

```
99 88 3 FRED com!
```

When array references are compiled into colon definitions (or type operations) a speed optimization is available by the use of VAL[. The following definitions both do the same thing :

```
: TEST1     5 FRED com@ ;
: TEST2     VAL[ 5 ] FRED com@ ;
```

However the second version will run at least 30% faster due to the use of VAL[, which tells the compiler that the index is a known constant value, and so permits full early binding. By inspecting the source code for ARRAY. COMP.OP and ARRAY.VAR, you will see the extra compiled code which is responsible for this difference in efficiency. VAL[is always optional, and must never be used when the index is not known at compile-time, as for example in

```
: TEST3    10 0 DO I FRED com@ LOOP ;
```

VAL[will work with a variable as index, but the result will not be normally what is required, as it will take the value of the variable at compile-time rather than run-time :

```
: TEST4    VAL[ X @ ] FRED ;
```

The initialization mechanism we devised using INITIALIZE will not work with ARRAY-OF as the latter does not use MAKE.INSTANCE to create instances. An analogous but more complicated mechanism could be incorporated into MAKE.ARRAY, but it is easier and safer to initialize array instances explicitly with a colon definition :

```
: INITFRED    10 0 DO I FRED INIT LOOP ;
```

Of course arrays which are instance variables in a type can be initialized in the INIT operation for the type just like any other variable. And by using ALLOTZ we have at least guaranteed that all array elements are initialized to zeroes.

Regrettably there is no way to test for the second major error condition (index out of bounds being the first), namely omitting to supply an index value at all. This is so because in the compiling state it is quite permissible for there to be no index on the stack at compile-time. A test for "no index" during interpretation only is possible, but of little value. This weakness is particularly unfortunate because most Forth systems are intrinsically insecure with regard to stack underflow (try DUP on an empty stack, or ROT with one item on the stack, in your system). This means that an attempt to perform an operation on an unindexed array, for example :

```
FRED com@
```

will probably "succeed" in the sense that it will return a spurious value, as a phantom index will be dredged up by stack underflow. Vigilance on the part of the programmer is the only way to avoid such unindexed array operations.

A Discrete Simulation Example

With the addition of arrays, we have more or less reached the end of the enhancements which are both easy and useful. There are still many features that we could add to this abstract data typing scheme, if we were determined for example to emulate all the features of Ada. The most obvious omission is a proper form of generic types, that would permit the declaration of, say, GENSTACK as a stack of type ITEM, where ITEM could be replaced by any type during instantiation.

It is however hard to imagine two languages further apart than Forth and Ada, and a lynch mob might well be the reward for those foolish enough to try and bring them together! The scheme as it now stands, with ARRAY-OF, is extremely powerful, still quite small (under 1.5K of compiled code in a Z80 system) and most importantly, fully compatible with the Forth style of programming.

Once one becomes accustomed to an object-oriented style of Forth programming, many application areas become very much easier to handle. Discrete simulations are a good example, where the interactive nature of Forth combines superbly with the concept of objects having both attributes and behaviour. A discrete simulation is a model of a system in which discrete (i.e. separated in time or space) events occur, as opposed to a continuous simulation such as the trajectory of a projectile, in which all the variables vary in a continuous fashion.

To demonstrate the object-oriented style of programming, let us write a discrete simulation program using abstract data types. The problem I have chosen is one of the text-book classics, namely, to simulate queueing in a bank with a number of cashiers' windows.

Customers will arrive (singly) at random intervals and go to a random cashier's window. There they will queue if necessary until served. Each customer will take a finite but variable time to be served, which we will set to be random but constrained within limits. By playing around with the average time between arrivals and the average time to be served, we can investigate the way that queues build up, and by modifying the model and the customer's behaviour perhaps discover ways to reduce queueing.

It is normal for such simulations to be multitasked, so that events can occur in a pseudo-parallel fashion. It is quite easy however to interleave separate tasks manually in high-level Forth, and we shall take this route; readers who have multitasking Forth systems may wish to modify the code accordingly. We assume that a pseudo-random number generator called RANDOM is available, which takes a number from the stack and returns a random number between (and including) zero and that number.

Two type declarations will be needed, a type CUSTOMER which records both the attributes and the behaviour of customers, and type CASHQ which is a queue of customers. The bank will then be represented by an array of CASHQs, with an element for each cashier.

First we must decide what attributes a customer is to have (to allow for later expansion, we may decide to record more information than will be initially used). Let us choose to record for each customer the duration of their intended transaction with the cashier, their time of arrival in the bank, and a status flag which records whether or not they have been served.

The behaviour of customers will be to enter the bank and join a queue, to carry out a transaction with the cashier when they reach the head of the queue, and to reply to the query "are you finished?"

This can be captured by the type declaration :

```
TYPE> CUSTOMER
  2 VAR TRANSACT.TIME              -- duration of desired transaction.
  2 VAR ENTRY.TIME                 -- time bank entered.
  2 VAR DONE                       -- flag for completion of transaction.
  10 CONSTANT TDELAY               -- upper limit for transaction time.
OPS>

  -- new customer joins queue
  : JOIN       TDELAY RANDOM 5 +     -- a number between 5 and TDELAY+5.
               TRANSACT.TIME !       -- initialize duration.
               FALSE DONE !    ;     -- not served yet.

  -- carry out one time slice of a transaction
  : TRANSACT   -1 TRANSACT.TIME +!   -- decrement by one time unit.
               TRANSACT.TIME @
               IF   FALSE DONE !     -- still being served.
               ELSE TRUE DONE !      -- finished.
               ENDIF ;

  -- test for completion of transaction
  : DONE?      DONE @ ;                          ( --- flag)

ENDTYPE> CUSTOMER
```

The CASHQ is a queue created according to textbook principles. A queue data structure is a collection of items to which new items can be added (called "enqueuing") and from which old items can be removed (called "dequeuing"); unlike a stack, however, the first item into a queue is also the first to be removed. A queue is often called a FIFO (First In/ First Out) structure, whereas a stack is a LIFO (Last In / First Out) structure. Like a stack, a queue requires a pointer to the head of the queue (the place from where an item can be removed) but unlike a stack, it also requires a pointer to its tail, where new items are added. We commonly record the length of the queue instead of a tail pointer; the tail can then be found by adding the length to the head pointer. An array is suitable for holding queues whose maximum length can be specified in advance :

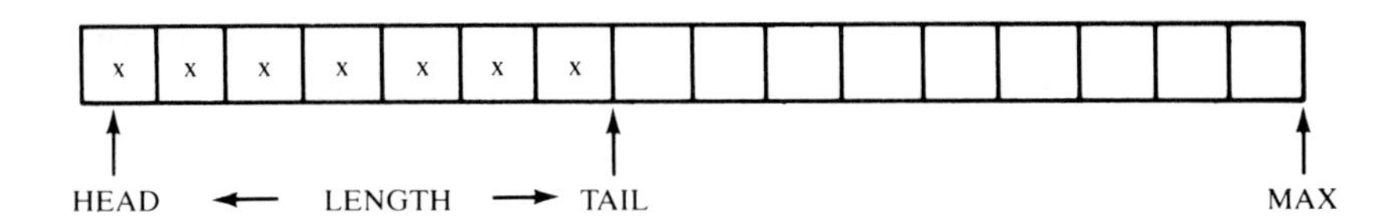

An array of customers will form the body of our queue, and for the sake of efficiency we shall use a *moving* head pointer, rather than fixing the head at the first element (which would mean that all the elements would have to be moved up one place after a dequeue). The dequeue operation must increment the head pointer after removing an item.:

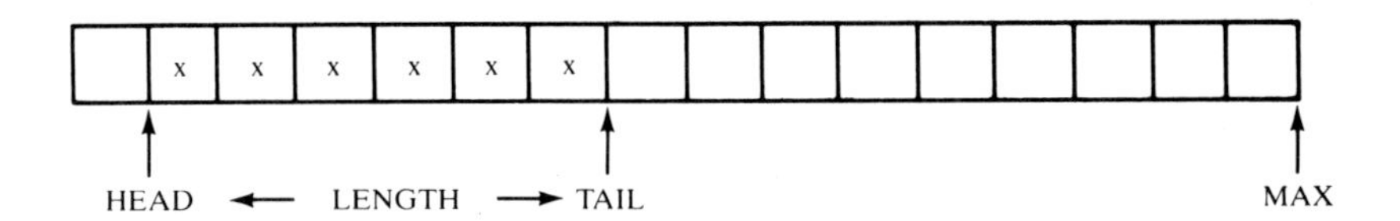

As more items are added to the tail, this means that eventually the tail pointer can run up against the maximum, even though there are free slots before the head. In this case the tail pointer is allowed to "wrap around" to the beginning again, and this form of queue is often for this reason called a *circular buffer*. Such a queue is only full when the tail pointer meets the head pointer :

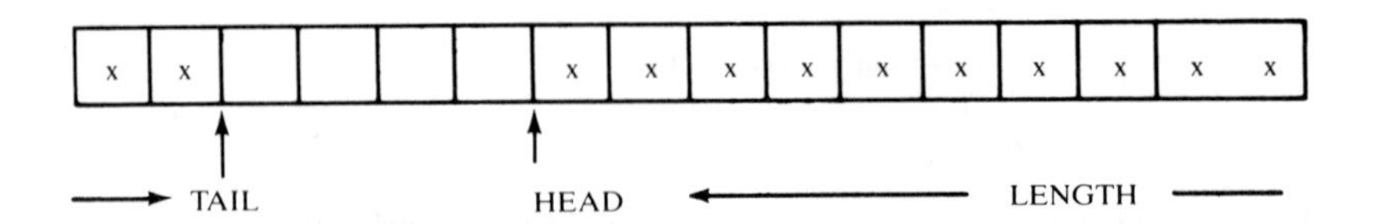

Though this appears rather difficult to visualize, it is very easy to handle in practice. We simply perform all arithmetic on the pointers MOD (size of the array).

CASHQ will have an enqueue operation which merely performs the JOIN operation on the first free customer in the queue. In other words we are not actually creating any new instances of CUSTOMER at all, but "faking" it by reusing the queue elements as if they were new objects.

Similarly the dequeue operation merely moves the head pointer, and does not return any object so our customers vanish into thin air after they have been served. This is quite adequate in a computer model, if not in a real bank! In more sophisticated models one might wish to "interview" customers after they have been served, in which case the dequeue operation would include such an interview.

Here is the declaration of CASHQ :

```
TYPE> CASHQ
  2 VAR HEAD
  2 VAR LEN
  20 CONSTANT MAXQ
  MAXQ ARRAY-OF CUSTOMER QBODY
OPS>
  : FULL?    LEN @  MAXQ = ;          -- is queue full?    ( --- flag)
  : EMPTY?   LEN @  0 =  ;            -- is queue empty?   ( --- flag)

  : DQ       EMPTY? NOT IF HEAD @ 1+ MAXQ MOD HEAD !      -- bump head
                        -1 LEN +!                          -- decr. length
                      ENDIF ;

  : NQ       FULL? NOT IF HEAD @ LEN @ + MAXQ MOD          -- get tail
                        QBODY JOIN                         -- add customer
                        1 LEN +!                           -- bump length
                      ENDIF ;

  : SERVE    HEAD @ DUP QBODY TRANSACT                   -- serve head of queue
             QBODY DONE? IF  SELF DQ  ENDIF ;            -- vanish if done!

  : SHOWQ    EMPTY? NOT IF  LEN @ 0 DO ." #" LOOP        -- print queue
                      ENDIF ;

  : INIT     0 HEAD ! 0 LEN ! ;          -- initialize the queue pointers

ENDTYPE> CASHQ
```

Actually we will only need to see the operations SERVE, NQ and SHOWQ, but FULL?, EMPTY? and DQ are useful during debugging. When debugging is complete we could move OPS> to just after DQ to hide the rest of the operations. INIT is included for good form, though we shall in fact rely on the initialization performed by ALLOTZ.

With these two types we can now produce the main program loop. The "bank" will be an array of CASHQs. Timing is simulated by incrementing the variable CLOCK at the beginning of the loop, so that each pass through the loop represents one time slice. A variable NEXT.CUSTOMER is initialized with a random number, representing the delay before a new customer enters the bank. A test is made each time-slice to see if a new customer is required. Following this, an inner loop services the customers at the heads of all the queues, using up one time-slice of their transaction time, and then prints out a character-graphic display of the bank.

```
-- working variables.

   VARIABLE CLOCK     VARIABLE NEXT.CUSTOMER     VARIABLE LAST.CUSTOMER

-- maximum delay before new customer enters.

     20 CONSTANT CDELAY

-- is a new customer due?                                      ( --- flag )

     : TIME.FOR.NEXT?   CLOCK @ LAST.CUSTOMER @ -   NEXT.CUSTOMER @ >  ;

-- set up the "bank".

     10 CONSTANT BSIZE

     BSIZE ARRAY-OF CASHQ CASHIER

-- main loop.

    : QSIM      0 CLOCK !                                -- initializations
                0 LAST.CUSTOMER !
                CDELAY RANDOM  NEXT.CUSTOMER !
                BEGIN
                  1 CLOCK +!                             -- advance one tick
                  TIME.FOR.NEXT?
                    IF  BSIZE RANDOM CASHIER NQ          -- new customer at random
                        CLOCK @ LAST.CUSTOMER !          -- record time of entry
                        CDELAY RANDOM NEXT.CUSTOMER !    -- time till next
                    ENDIF
                  BSIZE 0 DO  I CASHIER SERVE            -- serve head customers
                              I . I CASHIER SHOWQ        -- display queues
                          LOOP
                ?TERMINAL UNTIL ;
```

For serious use it would be better to make TDELAY and CDELAY accessible interactively so that the parameters can be varied during a run. The display part of this word is intentionally left very crude, merely producing a scrolling list of the queues, e.g. :

```
0 ####
1 ##
2
3 #
4
5 ##
6
7 ###
8 #
9 #
```

Given a cursor addressable terminal or PC display, preferably with a "delete-line" function, this can be simply converted to a non-scrolling, animated display, but such code is too machine-specific to consider here.

Some comments on the program: notice that the ENTRY.TIME field of the customer has not been used. It could be used in enhanced versions of the program which ascertain how long customers are kept waiting. Note also the questionable assumption that customers join a queue at random when entering the bank. It might be more realistic to make customers join the shortest queue available instead (though the above model would be fairly realistic if each queue were for a different kind of transaction).

These sort of details are not really important. What is important is the ease and compactness of the solution using ARRAY-OF and typed objects.

The resulting code is quite transparent, with little distracting implementation detail visible. The types we created actually model the real objects quite well.

Most importantly of all, the main program is completely isolated from the implementation details of CUSTOMER and CASHQ. If we decided that a linked list implementation of the queues were preferable to an array, nothing in the main program need be altered, so long as NQ, SERVE and SHOWQ do their jobs as before. The abstract types provide us with a "kit" of parts from which to build programs, without worrying about how they are implemented. If this became a large program, with several programmers involved, the task of ensuring correct interaction between parts of the program would be greatly eased since no one has to worry about clashes of variable names or illegal access to global data structures.

Error Reporting

Now that we have a taste of what object-oriented programming is going to be like, a potential problem area can be anticipated.

Since in typical programs we will be creating large numbers of objects, error reporting will need to be very precise. When an operation fails, either because it is not yet debugged or some value in its environment goes out of limits, how can we know which object the error affected? If there were hundreds or thousands of them this would not be a trivial problem.

It is, however, easily solved. We can make the error reporting in the system much more sophisticated than it currently is by providing one extra word called SELF.ID. This word merely causes the current object to print its name at the terminal. Since the PFA of the current object is always present on the top of OSTACK, SELF.ID could be simply defined as

```
: SELF.ID    OSTACK @ @ BODY> >NAME ID. ;
```

This definition includes the non-standard word ID. which prints the name of a Forth word, given its NFA. This word is in fact included in most commercially available systems, sometimes under another name such as .NAME. I cannot give a general implementation of the word for those who do not have it, as it depends crucially on the particulars of name storage in your system.

Taking our earlier stack example :

```
TYPE> STACK
  2      VAR STACKPTR
  50 2* VAR STACKBODY
  : INC    2 STACKPTR +! ;
  : DEC   -2 STACKPTR +! ;
OPS>
  : INIT     STACKBODY STACKPTR ! ;
  : EMPTY?   STACKPTR @ STACKBODY = ;                    ( --- flag)
  : PUSH     STACKPTR @ ! INC ;                          ( n ---    )
  : POP      EMPTY? NOT IF    DEC STACKPTR @ @           (    --- n )
                        ELSE ." stack empty in " SELF.ID
                        ENDIF ;
ENDTYPE> STACK
```

we can now make a stack announce its name when underflow occurs. In an application which used several stacks, this information would be valuable.

The word SELF.ID should be incorporated into all the error routines you write into operations, to announce the victim of the error. A problem exists for array elements though; the address of the *element*, rather than the array will be on OSTACK, and an element does not have a name field. Hence SELF.ID will print garbage. This will apply both to array objects, and to arrays which are instance variables in a type.

Working back from an element address to find that of its header is not an easy matter, but it can be done by brute force and ignorance. Noting that the code field of every ordinary object contains the same value, namely the address of the DOES> code in MAKE.INSTANCE, and similarly that the code field of every array object contains the same value, the address of the DOES> code for MAKE.ARRAY, we can initiate a backward search through memory for one of these values! The resulting address will be the code field of the nearest enclosing object, and from here SELF.ID can print its name. The most unpalatable part of this scheme is that absolute addresses need to be used (the DOES> code addresses), which will alter if the code is compiled at a different address or changed in size. However, if you plan to make types a permanent part of your Forth system, it is just about acceptable.

Firstly you must discover the relevant absolute addresses (Forth has no legal mechanism for locating DOES> code). To do this, create an object and an array of any type, and inspect their code field contents thus :

```
          COMPLEX FRED          ' FRED @ .    27767 ok

20 ARRAY-OF COMPLEX JIM         ' JIM @ .     28276 ok
```

Now SELF.ID can be rewritten as

```
: SELF.ID  OSTACK @ @ BODY>  DUP @ 27767 =
           IF   >NAME ID.
           ELSE BEGIN  1- DUP   @ DUP  28276 = SWAP 27767 = OR
                UNTIL >NAME ID.
           ENDIF ;
```

If this definition causes you involuntary shudders of disgust, you are not alone! The only justification is that it works.

We can also make FIND.OP much smarter in its error reporting, by noticing that in every case where FIND.OP is used, the PFA of its object is already on the parameter stack, beneath the key. Hence

```
: FINDOP  BL WORD SWAP UNLOCK FIND LOCK                ( addr key --- CFA )
          0= IF COUNT TYPE ." not recognized by "
                BODY> >NAME ID. ABORT
             ENDIF ;
```

Now if we request an illegal operation, we get a message like :

```
FRED ##       ## not recognized by FRED
ok
```

This mechanism can be partially implemented by Forth-79 users; the name of the object can be printed but not that of the operator (since no string is returned by 79-FIND).

Deferred Binding

A further problem which deserves some consideration is deferred or delayed binding. We have gone to great lengths to make the abstract data types early binding in the name of efficiency. However there are circumstances, particularly in advanced applications, where we should like to put off binding until run-time.

Let us be quite clear as to what we are referring here. Objects as currently implemented are early-bound in the sense that the applied operations are looked up at compile-time and compiled as normal Forth code. This means in effect that the type of an object must be known at compile-time. In fact, with the exception of elements in an ARRAY-OF, the actual *name* of the object needs to be known at compile-time, because our implementation makes object names into executable Forth words which take an operation name from the following input stream.

To exploit the power of abstract data types to the full, we should like to be able to pass pointers to objects as parameters, and to return pointers to objects as results from an operation; in the next chapter on linked lists we shall see how useful it can be to maintain lists of pointers to objects. It is easy enough to produce a pointer to an object. For example,

```
COMPLEX FRED          ' FRED >BODY
```

leaves the PFA of FRED on the stack. This could then be stored in a variable or passed on the stack to another Forth word. But when we write a word which takes the PFA of an object from the stack and applies an operation to it, we are in effect saying that neither the type nor the name of that object can be known until run-time. Such deferred binding allows the freedom to write words which can work on more than one type; so long as a group of types all share an operation with the same name, then a deferred binding word could apply that operation to objects of any one of those types. Objects of those types could be freely mixed in an array or a list. This freedom has many applications; for example, if the operation in question were called *display* we could write graphics programs of great generality, working on a whole variety of objects each of which has its own notion of what to *display* means.

There is nothing in our implementation which forbids such late binding in principle; indeed we already use it in interpretive mode (via DO.OP).

The word APPLY takes the PFA of an object and the address of a string, representing the name of an operation, from the stack and applies the operation to the object :

```
: APPLY    OVER @                              -- get key  ( pfa string ---    )
           UNLOCK FIND LOCK                    -- look up op
           IF  SWAP OPUSH EXECUTE OPOP
           ELSE COUNT TYPE ."  failed to bind with "
                BODY> >NAME ID.
           ENDIF ;
```

It is used as in this sequence :

```
VARIABLE ANY-OBJECT

: TESTWORD  ANY-OBJECT @  " com@" APPLY ;

COMPLEX FRED
' FRED >BODY  ANY-OBJECT !

TESTWORD
```

Again we need that non-standard word " to define string literals. Note also that APPLY cannot be used on array elements, because the required index calculations are not performed in APPLY. The idea of writing a separate ARRAY-APPLY is not very attractive.

Forth-79 will not support APPLY because 79-FIND demands to have its argument in the input stream at run-time; APPLY could therefore only be an immediate word, which negates the whole point of its existence.

As an adjunct to the use of APPLY, we could take a further leaf out of Smalltalk-80's book, and define the word ^ which, when used in an operation, returns the current object's address on the stack. The definition is just

```
OSTACK @ @ ;
```

and it is traditionally used as the last word in an operation. By using ^ the PFA of an object can be obtained as the result of an operation on it, without the need to use '. We could for example redefine COMPLEX:

```
TYPE> COMPLEX
  2 VAR REAL
  2 VAR IMAG
OPS>
  : INIT  0 REAL ! 0 IMAG ! ;
  : com!  IMAG ! REAL ! ^ ;        ( n n --- PFA )
  : com@  REAL @ IMAG @ ;          ( --- n n )
ENDTYPE> COMPLEX
```

Now whenever a store operation is performed on a COMPLEX object, it returns its PFA. Then APPLY could be used in the operations of another type to enable objects of type COMPLEX to be passed as arguments to those of the second type :

```
TYPE> TESTTYPE
  2 VAR SUM
OPS>
 : grab  " com@" APPLY  +  SUM +! ;    ( complex ---   )
ENDTYPE> TESTTYPE

COMPLEX FRED  TESTTYPE JIM

23 45 FRED com! JIM grab
```

The programming tricks enabled by this development are legion, and would be subject matter enough for another, very large, book in itself.

The use of deferred binding raises some delicate issues of semantics. The special message "failed to bind with" is deliberately introduced in place of "unrecognized operator". This is because failure to bind in a late binding implementation is an error of a rather different kind from merely applying an operator incorrectly. If we fully exploit the freedom to mix types provided by late binding, it may not even be an error at all, but rather information that needs to be acted upon. For this reason we have not followed the message with the customary ABORT; it may be more appropriate for processing to continue, perhaps to try a different possible binding.

When writing complex list processing programs using objects of mixed types, binding failure may have to be accepted as an ever present possibility rather than a fatal error. This situation is likely to arise in many Artificial Intelligence applications.

References

Pountain, R.J. (1986). "Object oriented extensions to Forth". *Journal of Forth Applications and Research,* Vol.3 No.3, pg 51. Institute of Applied Forth Research, Rochester.

Schorre, D.V. (1980). Forth Dimensions, Vol.2 No.5, pg 132. Forth Interest Group, San Jose.

Suggestions for Further Reading

Duff, C. and Iverson, N.D. (1984). "Forth meets Smalltalk". *Journal of Forth Applications and Research,* Vol.2 No.3. Institute of Applied Forth Research, Rochester.

Ghezzi, C. and Jazayeri, M. (1982). "Programming Language Concepts". Wiley.

Goldberg, A. and Robson, D. (1983). "Smalltalk-80: the Language and its Implementation". Addison-Wesley.

3 Lists

The two kinds of structured data which were discussed in the last two sections, namely records and abstract data types, share a common feature. They both allocate memory in a *static* way. That is, the size of the data structure is determined at compile-time and then permanently fixed; the size of an object cannot be changed at run-time.

In many circumstances it is useful to have access to data structures whose size can be altered at run-time, and such structures are called *dynamic* data structures. The stack and the queue are both data structures which have a dynamic aspect, in that their number of items can be varied at run-time. However there are strict limits on this variability. Both a stack and a queue may only be varied by a single item at a time. Moreover there is a definite order to the variability; in a stack for instance, only the last entered object can be removed, and all objects below the stack top must remain fixed in place until uncovered by successive POP operations. Furthermore, we chose to implement both the stack and the queue using an array as the underlying physical representation, which meant that the *maximum* size was fixed at compile-time, to be the size of the array used.

A far more flexible dynamic data structure is the *linked list*. A linked list consists of a series of data storage areas which, rather than merely being placed next to each other in memory, as in an array, contain pointers to each other. In a singly linked list, each list element contains the address of the next in the list, and so the list can be traversed (in one direction only) by following the pointers :

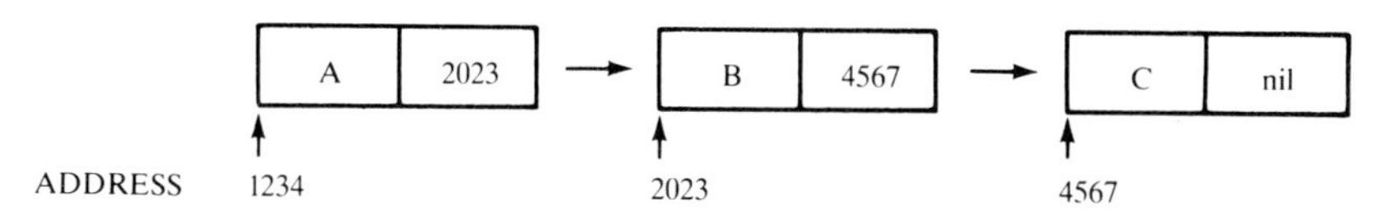

The elements of a linked list are more properly called *nodes*; a node consists of a data storage area or information field, and a *next node* field which contains the pointer. The end of a list is marked by putting some value which cannot be a legal element address, traditionally called *nil*, into the next node field. Physically nil is often merely represented by a zero value. It can be seen at once that one disadvantage of a linked list compared to an array is that more memory is consumed; two bytes per node in a 16-bit environment like Forth.

Forth is no stranger to linked lists for the most important data structure in Forth, the dictionary, is itself a singly linked list. Every word in the dictionary contains a pointer, the link field, which points to the previous word, and it is by following these pointers that Forth conducts dictionary searches (the nil value is 0). It is for this reason that the dictionary cannot be traversed backwards in Forth; a word contains no information about the location of the word which *follows* it in the dictionary. In the last section on abstract data types we were performing linked list operations, without naming them as such, to create private dictionaries.

The principle virtue of the linked list is that it can be increased or decreased in size dynamically. Elements can be added to a list (at any position) merely by altering pointer values, without moving any data at all. In contrast, to add an element to the middle of an array requires moving all the following elements to make room. The elements of a list may lie anywhere in memory, and do not have to be contiguous, which means that so long as there is free memory available somewhere, a new element can be added :

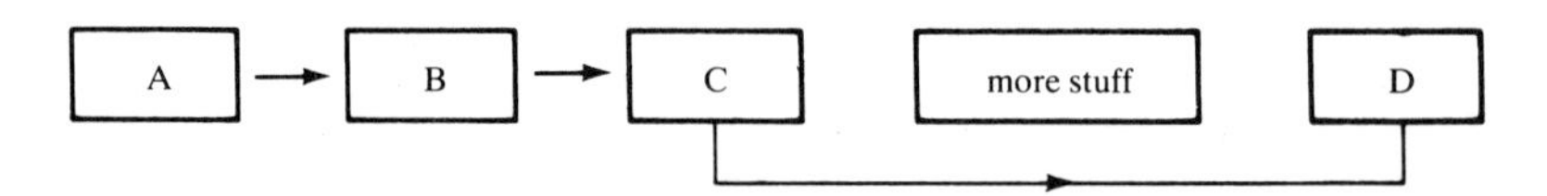

Removing an element from a linked list is equally easy, since the pointer of the preceding element merely needs to be redirected to point to the following element :

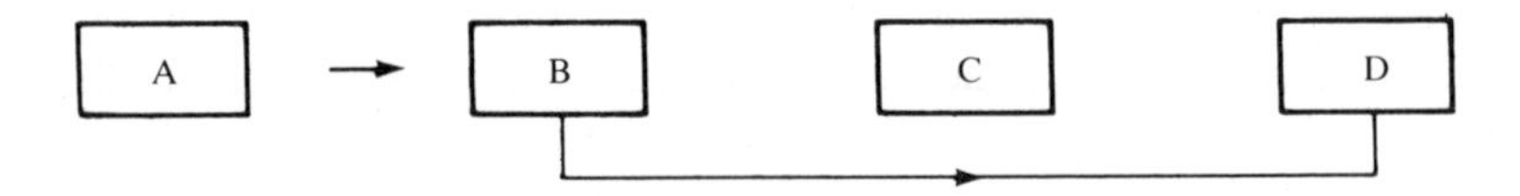

The element C has now been removed from the list. However C still occupies memory, and this memory cannot be used for anything else as the pointer to it has been lost. This constitutes the main disadvantage of linked lists, namely the possibility of *memory fragmentation*. Recovering such pieces of dead memory for reuse is called *garbage collection* and is a vast subject in itself.

The simplest case of the linked list is a singly linked list in which all the elements are the same size. In this special case garbage collection can be performed in a simple and elegant way. Let us see how such lists can be created in Forth in the simple case where the storage area is just a 16-bit cell.

Singly Linked Lists of Constant Sized Elements

The basic component of our Forth lists will be a four byte node, two bytes of which are the next node pointer, and two bytes data. The pointer will be placed in the lower addresses of each node and will point to the next pointer field. This allows maximum speed when traversing the list, which can be done by successive fetches with @.

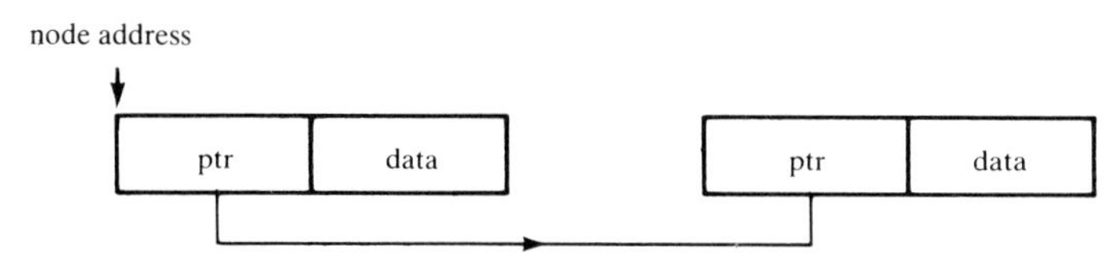

As a first attempt, we can create such nodes by simply ALLOTing four bytes in the dictionary, and returning the address of the new node :

```
: NEWNODE   HERE 4 ALLOT ;                  ( --- node )
```

Each list requires a way into it, in the form of a pointer to its first node; this is called the list header. The header can be an ordinary Forth word so that lists become named objects like variables or constants. When a list header is initially created it will be empty, and so by convention will contain nil :

```
0 CONSTANT NIL
: NEWLIST    CREATE NIL , ;      ( +++ )
```

When a list header is executed it will return its PFA, which contains the pointer. We shall denote such an address by "list" in the following stack content annotations, e.g. (list --- node).

Now two words are required to add a new node to the list, and to remove a node from the list. There are many possible ways to perform these operations. New nodes can be added or removed from the front of a list, the end of a list or to any arbitrary position in the middle of a list.

At first sight adding to the end of the list seems attractive as it, means merely replacing the nil pointer with a new node address :

(a)

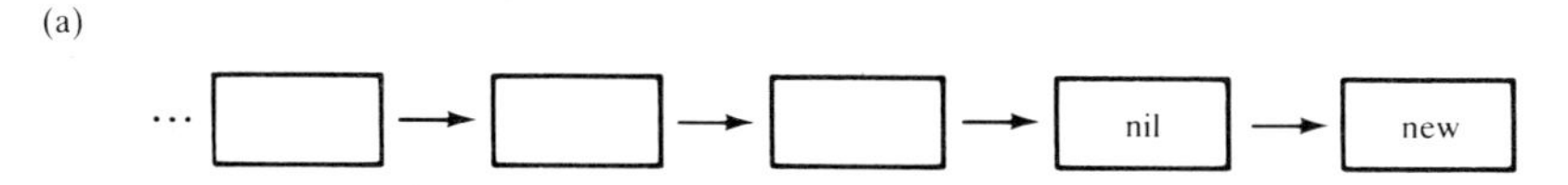

(b)

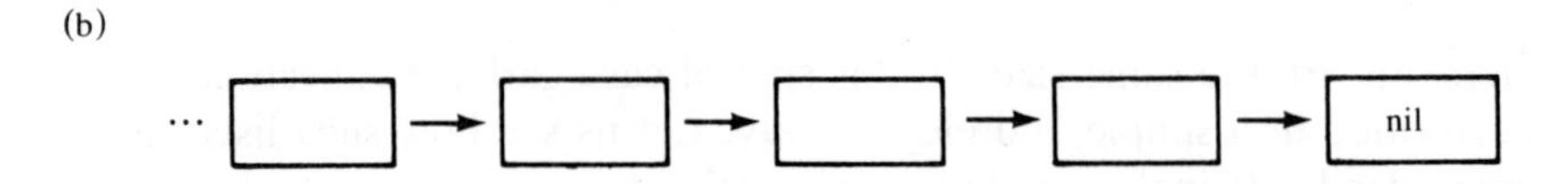

In fact this turns out to be a thoroughly bad idea, because it involves traversing the whole list to find out the address of the last node. Thus the time taken to add a new node (or remove one) rises in proportion to the size of the list, and can become very large for long lists. Since a singly linked list is accessed from one end, by following the successive pointers, it makes much more sense to add new nodes to the front :

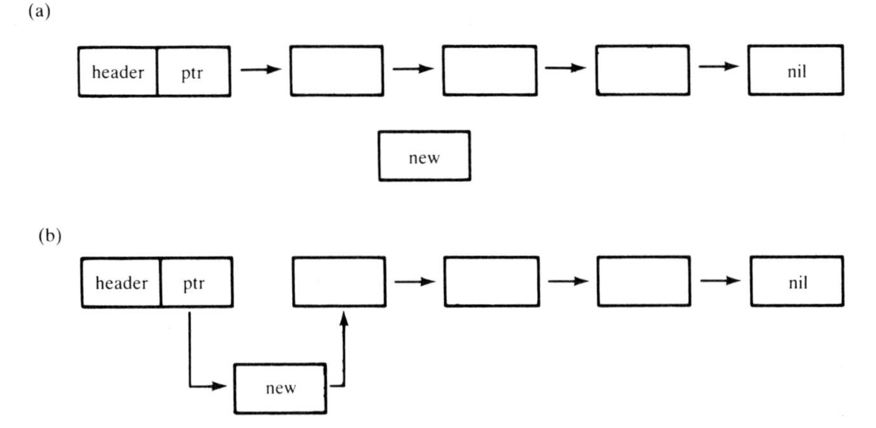

Removing a node reverses this process, i.e. goes from (b) to (a). The time taken to add or remove a node is now short, and unrelated to the length of the list. Words to perform these two operations are :

```
: INSERT   OVER @          -- get first node address   ( list node --- )
           OVER !          -- store in new node
           SWAP ! ;        -- store new node address in header


: REMOVE   DUP @ DUP       -- get first node address   ( list --- node )
           IF DUP @        -- if not empty, get second node address
              ROT !        -- and store in header
           ELSE SWAP DROP  -- otherwise return nil
           ENDIF ;
```

REMOVE incorporates a test for the empty list (whose header contains the nil pointer) and returns nil, which can be used in security tests by outer definitions. Also note that REMOVE works properly on a list with only one node; the second node address will just be nil.

The set of definitions we now have is sufficient to create lists of nodes, but there are no words which actually store data into the nodes. We can now write these in terms of INSERT and REMOVE. This factorization (separating node manipulation from data manipulation) will prove very useful when we go on to more sophisticated schemes. Let us gather all the code together :

```
0 CONSTANT NIL

: NEWLIST    CREATE NIL , ;                                    ( +++ )

: NEWNODE    HERE 4 ALLOT ;                                    (  --- node )

: INSERT     OVER @ OVER ! SWAP ! ;                            ( list node ---  )

: REMOVE     DUP @   DUP IF    DUP @ ROT !                     ( list --- node )
                          ELSE SWAP DROP
                          ENDIF ;

: PUSH     NEWNODE               -- make a new node      ( value list ---  )
           ROT OVER 2+ !         -- store value in it...
           INSERT ;              -- ...and add to list

: POP      REMOVE ?DUP           -- remove first node    ( list --- value )
           IF 2+ @               -- fetch its data
           ELSE ." empty list " ABORT
           ENDIF ;
```

POP uses the fact that REMOVE returns the nil end of list pointer to prevent popping from an empty list. No such test for overflow is required in PUSH, because Forth itself will handle the only case of failure of NEWNODE, namely a full dictionary.

We have chosen to use the names PUSH and POP because the action of adding and removing items from the front of the list resembles the action of a stack. "Resembles" is perhaps too weak a word; this *is* a stack. The linked list can be used as an alternative to the array as a method of implementing a stack.

The syntax for using the above definitions is

```
NEWLIST FRED
2 FRED PUSH
FRED POP
```

When experimenting with lists it is useful to have a word which prints out all the data values held in a list. This can be easily written as

```
: .ALL    BEGIN @ ?DUP          -- fetch pointers while not nil   ( list --- )
          WHILE DUP 2+ @ .      -- print data
          REPEAT ;
```

The above naive scheme has one great strength and one great weakness, both of which lie in the area of memory allocation. NEWNODE has the charming property of returning a new node as long as there is space left in the dictionary; the memory management is thus left up to Forth itself and no prior decisions about stack size need to be made. Unfortunately POP has a flaw which more than cancels this charm; every time a node is removed from the list, the memory it contains is lost to use forever since there is no longer any pointer to it. After sufficient stack operations, all of the available memory will become useless!

The situation is easily remedied however. All that is necessary is to organize free memory itself into a list, and to return nodes to this list rather than discarding them.

The Free List

To manage node memory effectively, we must create a list called FREELIST which contains all the empty nodes not currently forming part of any other list; it forms a pool of available nodes. This means that we are back to the situation of deciding how much memory is to be made available to the lists; it is not practical to put all the free dictionary space into the pool as this would prevent us from compiling any colon definitions.

The position is still superior to that of an array, however. All the lists we create can share the same free list, and as long as there are free nodes in the pool, any list can be grown. When several data structures share the same array however, it is possible for one to run out of space while the others have plenty of space left, as in

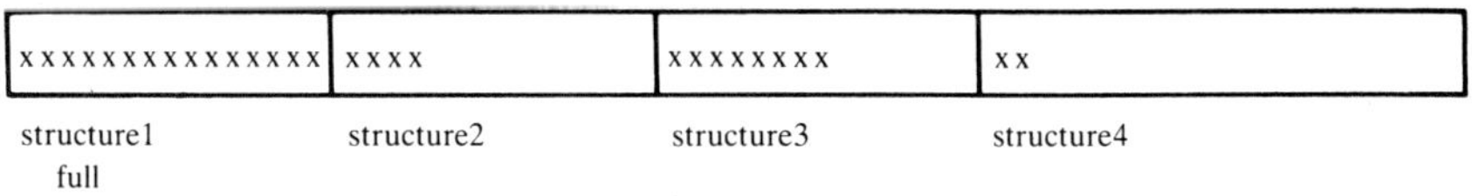

xxxxxxxxxxxxxxx	xxxx	xxxxxxxx	xx
structure1 full	structure2	structure3	structure4

To release any of the spare space for structure1 would involve moving array elements, and thus would be a very costly business. When using arrays, at most two structures can share a common pool of memory, by growing from opposite ends of the array; this is the way Forth itself shares memory between the dictionary and the parameter stack :

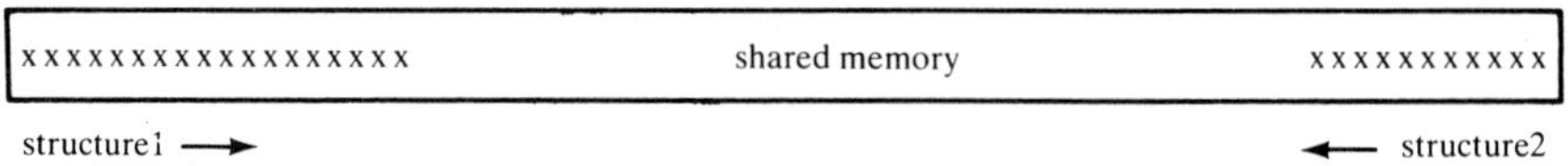

xxxxxxxxxxxxxxxxxxx	shared memory	xxxxxxxxxxx
structure1 →		← structure2

By contrast, using linked lists with a free list, any number of lists may equitably share the same pool of memory.

To set up a free list, all that is required is to allocate a certain amount of space and structure it as a list, with linked pointers. The data field contents are of no consequence.

```
: ALLOCATE   CREATE              -- make a header            ( size ---  )
             HERE 2+ ,           -- pointer to first node
             1 DO
                 HERE 4 + ,      -- link to next node
                 2 ALLOT         -- data field
               LOOP
             NIL , 2 ALLOT ;     -- make last node
```

Now a free list of, say, 4000 nodes can be created by

```
4000 ALLOCATE FREELIST
```

Very little of the previous code needs to be changed. LIST, INSERT, and REMOVE stay exactly as they were. NEWNODE of course needs to change fundamentally. Let us change its name to GETNODE, to reflect the fact that the nodes it gets are not necessarily new, and may well be second-hand! Its definition is trivial :

```
: GETNODE   FREELIST REMOVE ;                ( --- node )
```

GETNODE needs a partner called FREENODE, which hands back a node that is finished with to the free list :

```
: FREENODE  FREELIST SWAP INSERT ;         ( node ---   )
```

PUSH needs to have GETNODE replace NEWNODE and now also needs to test for the case of an empty free list. REMOVE tests automatically for an empty free list (which is after all a list like any other) :

```
: PUSH  GETNODE ?DUP                                   ( value list ---  )
        IF  ROT OVER 2+ ! INSERT
        ELSE ." no free space " ABORT
        ENDIF ;
```

POP must be redefined :

```
: POP  REMOVE ?DUP                                  ( list --- value )
       IF   DUP FREENODE                 -- recover unwanted node
            2+ @                          -- extract value
       ELSE ." empty list " ABORT
       ENDIF ;
```

Instead of squandering the space removed, POP now returns it to the free list where it can be "recycled" when a new node is needed.

Now any number of lists can grow and shrink at will, sharing the same pool of memory. The only restriction is that when the free list becomes empty, no more growth is possible.

Although this system manages the memory used and released by lists as they grow and shrink, there is as yet no operation to reclaim all the space of a list which is no longer needed. It is this operation, rather than the individual recovery of removed nodes, which is normally referred to as garbage collection. In sophisticated applications it is desirable for garbage collection to be performed automatically once the free space falls below a predetermined level. This requires complex algorithms to determine which lists are still in use by the program and which are now redundant; one such technique called "reference counting" involves keeping a tally of how many objects contain references to each list and killing those lists whose count falls to zero.

However we shall write a simple word KILL which is manually applied to lists no longer needed by the program, rather in the way that FORGET is used on Forth words :

```
: KILL    BEGIN DUP REMOVE ?DUP                        ( list ---  )
          WHILE  FREENODE
          REPEAT DROP ;
```

Note however that KILL cannot remove the list header itself, which is a normal Forth word and may or may not be safely removable by FORGET.

To recap, here is the new code for list manipulation with a free list :

```
-- constants

      0    CONSTANT NIL
      4000 CONSTANT MAXNODES

-- create a new list header

      : NEWLIST   CREATE NIL , ;                          ( +++ )

-- create the free list

      : ALLOCATE    CREATE                                ( size ---  )
                    HERE 2+ ,
                    1 DO
                        HERE 4 + ,
                        2 ALLOT
                      LOOP
                    NIL , 2 ALLOT ;

      MAXNODES ALLOCATE FREELIST      -- set up node space

-- node manipulation words

      : INSERT     OVER @ OVER ! SWAP ! ;                   ( list node ---  )

      : REMOVE     DUP @  DUP IF DUP @ ROT !  ENDIF ;       ( list --- node)

      : GETNODE    FREELIST REMOVE ;                        ( --- node )

      : FREENODE   FREELIST SWAP INSERT ;                   ( node ---  )

-- data manipulation words

      : PUSH  GETNODE ?DUP              -- add value to list ( value list ---  )
              IF  ROT OVER 2+ ! INSERT
              ELSE ." no free space " ABORT
              ENDIF ;

      : POP  REMOVE ?DUP              -- take value from list ( list --- value)
             IF   DUP FREENODE 2+ @
             ELSE ." empty list " ABORT
             ENDIF ;

      : .ALL    BEGIN @ ?DUP            -- print list contents ( list --- )
                WHILE DUP 2+ @ .
                REPEAT ;

      : KILL    BEGIN DUP REMOVE ?DUP  -- reclaim list space  ( list ---  )
                WHILE  FREENODE
                REPEAT 2DROP ;
```

List Constants

It would be very useful to have a notation for describing list constants, that is lists of values which can be assigned to list variables. Examples of such lists could be the list 5 8 7 3, or A B C D. Using our existing words, such lists could only be assigned to a variable by repeatedly using the PUSH operation:

```
NEWLIST FOO
5 FOO PUSH   8 FOO PUSH   7 FOO PUSH   3 FOO PUSH
```

This is verbose and clumsy, and we should like to be able instead to describe a list constant and assign it to a variable in one operation.

A list constant needs to be delimited in some way, and many languages use brackets for this purpose, e.g. (5 8 7 3). Forth has already laid claim to the parentheses (), for comments, and to the brackets [] for switching compilation states. On most computer keyboards this leaves the braces {}, and we shall use these as delimiters. If your Forth system already uses these for something else, then some contrivance like ((can be substituted.

The strategy is very simple. List elements will be placed as values on the stack, and a word called SET will insert them all into the list, using the following syntax :

```
{ 5 8 7 3 } FOO SET
```

Clearly SET must be able to take a variable number of arguments from the stack, and it will be the responsibility of { and } to leave an item count which makes this possible. The definitions are not difficult :

```
VARIABLE ITEMS

: {  DEPTH ITEMS ! ;                   -- record list start on stack
: }  DEPTH ITEMS @ - ;                 -- leave count above items ( --- c )

: SET  DUP KILL                        -- empty the list  ( ...n c list --- )
       SWAP 0 DO  SWAP OVER PUSH       -- push the items
              LOOP  DROP ;
```

If the DUP KILL were omitted, then SET would become APPEND, and would add the list constant to the front of the variable's existing contents.

Using this notation, we can create lists of variables, constants or lists, as well as literal values :

```
NEWLIST FRED
VARIABLE TOM   VARIABLE DICK   VARIABLE  HARRY
.......
{ 1 2 3 TOM @  DICK @  HARRY @ } FRED SET
```

A list of lists can be created like this :

```
NEWLIST FRED   NEWLIST TOM   NEWLIST DICK   NEWLIST HARRY
.......
{ TOM DICK HARRY } FRED SET
```

POP returns a pointer to one of the lists, and so FRED POP POP would return the first element of TOM. It is essential when playing such games not to inadvertantly mix up objects of different types in a list; putting a literal number in among list pointers is a guarantee of disaster when you POP it! List processing languages such as Lisp are constructed around a type checking mechanism which allows them to handle lists of mixed objects with impunity, but to develop such a mechanism is a subject for another book. However we can, within limits, handle mixed type lists using the abstract data types of the last chapter, as we shall see below.

Using Linked Lists

The simple form of linear, singly linked lists we have just developed can be used in place of arrays to implement a variety of data structures. The code above already constitutes a push-down stack for 16-bit integer values.

There would be little point in using the linked list implementation for a single stack, or even for two stacks, which would be more efficiently represented (in both memory and time terms) by an array. However if multiple stacks are required, the automatic memory management supplied by the free list becomes very attractive. Such multiple data structures can be dealt with in a tidy way by redefining NEWLIST. For example, a number of stacks could be handled as a single object by defining NEWLIST as an *array* of pointers rather than a single pointer :

```
: NEWLISTS    CREATE 0 DO NIL , LOOP          ( size ---    )
              DOES> SWAP 2* + ;

10 NEWLISTS STACKARRAY         -- set up 10 stacks

23 3 STACKARRAY PUSH           -- push 23 to the third stack

3 STACKARRAY POP               -- and so on.....
```

In principle it is easy to modify the list words to create lists with nodes larger than four bytes. By changing only ALLOCATE, PUSH and POP we can create lists with data fields of any size, to hold structured objects or records. In practice however it is more generally useful to keep the same 4-byte list nodes, and store *pointers* to complex data objects in their 16-bit data fields, just as we did with list pointers in the TOM, DICK, HARRY example above.

Records could be stacked like this :

```
NEWLIST RSTACK

-- definition of record type ADDREC

ADDREC TOM   ADDREC DICK   ADDREC HARRY

TOM   RSTACK PUSH
DICK  RSTACK PUSH
HARRY RSTACK PUSH
```

The value returned by popping this stack is the PFA of a record structure, which is precisely what is needed to access the record's fields :

```
RSTACK POP  .address1  COUNT TYPE
```

The abstract data types of the last chapter pose a slightly more complicated problem, as they are "active" data structures which take an operation name from the input stream, and they are not usually referred to simply by passing their PFA.

One solution is to introduce a LINK instance variable into each type which needs to be "listable", and then INCLUDE> a package of list processing operations similar to those defined above. The operation PUSH for example would cause an object to push itself onto a list whose address is on the stack. This solution suffers from the problem that the value fields of the nodes will be of different sizes for different types; this means that a single free list of nodes cannot be kept, and each type will require its own free list. This destroys much of the point of using lists.

The preferred solution is again to use the list code as it stands, to store PFAs of objects, and use the deferred binding word APPLY to execute them :

```
NEWLIST CSTACK

COMPLEX TOM   COMPLEX DICK   COMPLEX HARRY

' TOM >BODY  CSTACK PUSH  etc...

: COMPOP  CSTACK POP " com@" APPLY ;
```

This has the great advantage that all types are represented by similar 16-bit values, and so it becomes possible to have lists of mixed types but to still maintain a single free list. If, for example, we defined several different types, all of which had an operation called *display* which caused a screen display appropriate to the type, then we could create lists of mixed object types like this :

```
NEWLIST PRINTLIST

: SHOW   REMOVE DUP IF DUP FREENODE                    ( list --- flag)
                      2+ @ " display" APPLY
                      TRUE
                   ENDIF ;

: SHOW-ALL   BEGIN DUP SHOW WHILE REPEAT DROP ;        ( list ---   )

TYPEA A   TYPEB B  TYPEC C  etc....
' A >BODY PRINTLIST PUSH  etc...
PRINTLIST SHOW-ALL
```

As shown at the end of the last chapter, the word ^ could be used to get the PFAs of objects, without having to resort to ' >BODY.

Yet another possibility is to implement linked lists *as* an abstract data type. The linked list is just the sort of fragile mechanism which would benefit from the protection of information hiding; after all a single corrupted pointer can wreak havoc on the whole system. As an exercise, try using the TYPE> definitions of the last chapter to define types LISTNODE and LIST, with their appropriate operations. The free list would become an ARRAY-OF LISTNODE.

Linked lists are also highly suitable for implementing queues, and the same arguments apply *vis--vis* an array implementation to queues as to stacks. To make a linked list queue it is necessary to maintain two pointers in the list header, one to the first node (as above), and a second which points to the last node. In our array implementation in the last chapter, the queue *length* proved to be a suitable substitute for a tail pointer. This is not true for a linked list implementation, as using the length in this way would mean that the whole list has to be traversed (at great cost) every time the tail needs to be found.

We shall not proceed any further with a linked list queue at this point, because it will transpire later that a *circular* list is in fact far more appropriate for the purpose.

Other Operations and Enhancements

As mentioned previously, one advantage of a linked list over an array is that elements can be inserted or removed from the middle without having to move any data. The implementation given so far only inserts and removes from the front of a list. Or does it? In fact INSERT and REMOVE (and consequently PUSH and POP) work perfectly well on any node in a list, not just the header. Given the address of any node, they will insert or remove a node *immediately following* the given one. The only reason that they are confined to the front of the list at present is that the first node is the only one whose address we can find, since it is pointed to by the header.

By adding a word called TRAVERSE, we can find the address of any node in a list :

```
: TRAVERSE    0 DO @ LOOP ;
```

Now FRED n TRAVERSE will return the address of the n'th node in list FRED. We can add and remove elements like this :

```
29 FRED 4 TRAVERSE PUSH       -- insert 29 after the 4th element.
FRED 8 TRAVERSE POP           -- remove the element after the 8th element.
```

This simple-minded version of TRAVERSE is of course quite capable of running off the end of a list and wreaking havoc if so requested, so perhaps it should have some built-in safety :

```
: TRAVERSE    0 DO @ DUP 0= IF ." beyond list end " ABORT
                            ENDIF
                 LOOP ;
```

1 TRAVERSE returns, as expected, the first node of the list, but 0 TRAVERSE will run wild thanks to the behaviour of the Forth-83 DO...LOOP structure. For those of nervous disposition, a further test to reject traverse positions less than 1 could be added.

In some applications it will be convenient to maintain pointers into a list in addition to that in the list header. This can be done by defining a special version of PUSH that returns on the stack the node address supplied by GETNODE. This value can then be stored in a pointer variable, and used as an alternative entry point to the list :

```
: ^PUSH  GETNODE ?DUP                        ( value list --- node )
         IF DUP >R
            ROT OVER 2+ ! INSERT
            R>
         ELSE ." no free space " ABORT
         ENDIF ;

NEWLIST FRED     VARIABLE ^MARK

23 FRED PUSH etc. etc......

99 FRED ^PUSH ^MARK !
```

In this case there is no need to use TRAVERSE. Elements can be added or removed directly after the one pointed to by ^MARK by saying, for example,

```
33 ^MARK @ PUSH
```

while pushing to FRED adds them to the front of the list as normal. Obviously any number of nodes could be marked in this fashion. One could merely define PUSH to perform the ^PUSH action and drop the unwanted node value most of the time, but having the special version is neater and more comprehensible.

Note the restriction to accessing elements *after* a known node address. It is imposed by the singly linked nature of our lists; in particular it is quite impossible in principle to access any nodes *before* a given one.

If this presents a problem for a particular application, there are two solutions. One is to use doubly linked lists. This is an expensive solution in memory terms, as each node now contains two pointers, that is four bytes of overhead :

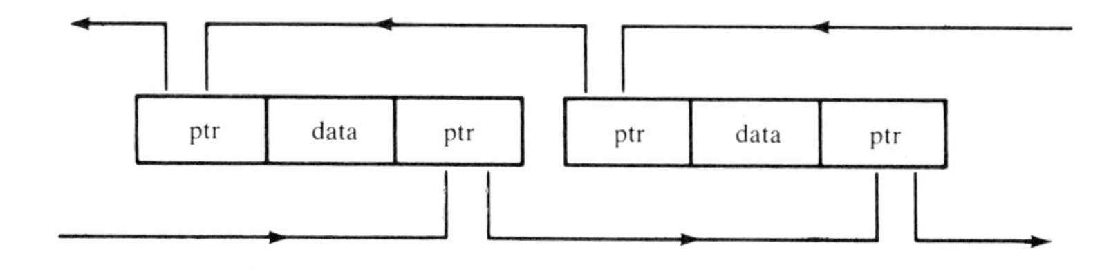

Such a list can be traversed in either direction, given a pointer to either end.

Another solution is to use a trick we have seen before, when implementing a queue in the last chapter. That is to make the list circular. In other words, instead of the last node containing a nil pointer, it points back to the first node in the list :

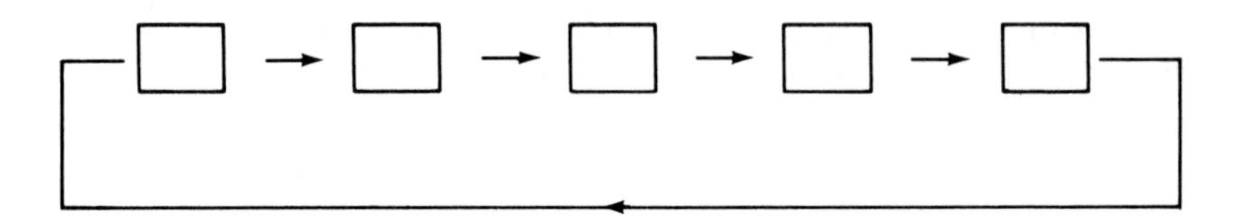

In such a list, any element can be reached from any other by traversing far enough around the circle. However if the list is large this can take time, so the two solutions offer a trade-off of memory for time.

In a circular list, the question arises as to what is the front and what the end. A useful convention is to say that the node pointed to by the header is the *last* in the list, and its following node is the first :

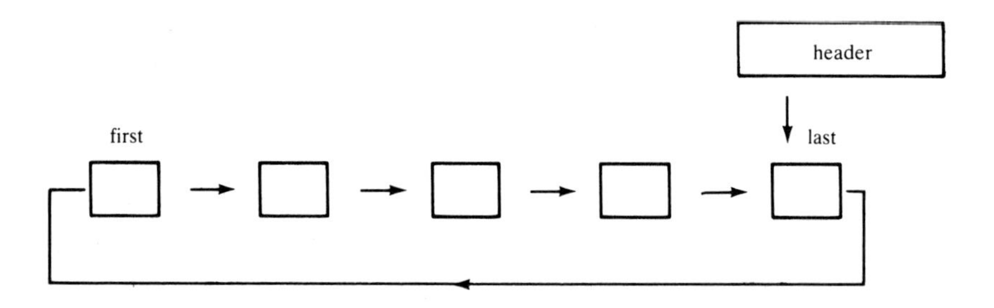

Another question arises over how one detects the end of such a list, since there is no nil value to test for. The answer is that it is not possible; either we must keep a record of the length of the list in its header (incremented by INSERT and decremented by REMOVE), or else use circular lists only for jobs which do not demand full traversal of the list.

A stack or queue is just such a case. In either case we are only interested in the empty condition, and that can be easily tested for by making the header, again by convention, contain a nil pointer. Moreover circular lists have properties which make them ideally suited for implementing queues.

The Circular List

The easiest way to modify our previous code to implement circular lists would be to merely make the last node point back to the header itself; this could be done by simply initializing the header with its own address instead of nil :

```
: NEWLIST   CREATE HERE , ;
```

All the other words would remain the same except for REMOVE, which would test for the empty list, not by looking for a nil pointer, but by looking for a pointer which is the same as the header address, and then returning zero.

However this is a bad strategy, because it makes the header part of the circular list, despite the fact that the header is not a normal node (for example, it cannot legally be given back to the free list). The effect is to destroy the property which makes the circular list so valuable, namely that the first element can be made to be the last element merely by advancing the header pointer by one place.

Instead we shall choose to do a little more work to keep the header as an external pointer into the list, as shown in the last diagram. The extra work is required because the nil pointer in a header, which marks an empty list, is now purely a convention, rather than being functional as it was before. This means that INSERT needs to test for an empty list, as inserting the first element now becomes a special case. REMOVE needs to test not only for an empty list, but also the list with only one element, for now a nil pointer must be specially created; the last node's pointer is no longer useable. There are three separate cases to consider :

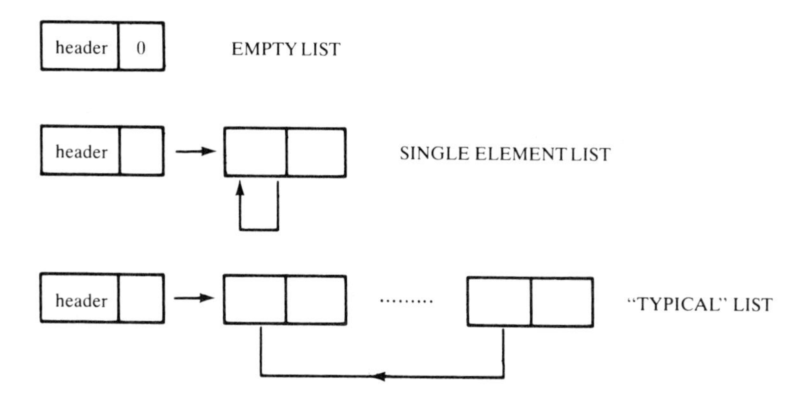

The new definitions of INSERT and REMOVE appear as follows :

```
: INSERT    OVER @ ?DUP          -- is list empty?       ( list node ---  )
            IF   2DUP @          -- get last node's pointer
                 SWAP !          -- put into new node
                 !               -- link old to new node
                 DROP            -- header address not needed
            ELSE DUP DUP !       -- make new node point to itself
                 SWAP !          -- make header point to it
            ENDIF ;
```

If you trace through the action of INSERT you will notice an important difference from our linear lists. The external pointer value does not change; though the newly added node becomes the *first* in the list, the header remains pointing to the last node.

```
: REMOVE   DUP @ DUP                -- is list empty?     ( list --- node)
           IF   DUP @ 2DUP =        -- does it have only one node?
                IF   DROP 0 ROT !   -- then make header pointer nil
                ELSE DUP  @ ROT !   -- else move first node pointer to last
                     SWAP DROP      -- header address not needed
                ENDIF
           ELSE SWAP DROP
           ENDIF ;                  -- return a node or nil
```

REMOVE exactly reverses the effect of INSERT, thus removing the first node of the list. Using these definitions, our old versions of PUSH and POP will work as before and the circular list becomes a stack. One further detail is necessary. We should make the free list into a circular list too, since the new INSERT and REMOVE expect to work on such a list :

```
: ALLOCATE    CREATE  HERE 2+ DUP ,     -- make header     ( size ---    )
                SWAP                    -- keep start address
                1 DO HERE 4 + ,         -- next field
                     2 ALLOT            -- info field
                  LOOP
                , 2 ALLOT ;             -- last node points to start
```

The words NEWLIST, GETNODE and FREENODE remain unchanged.

Remarkably little needs to be done to make the circular lists behave as queues rather than stacks. PUSH puts a new node at the beginning of the list, but leaves the header pointing to the last node. If we merely advance the header pointer by one node after a PUSH, then it will point to the newly added node, which then becomes the last node rather than the first. POP need not be altered at all; it will now remove the last item rather than the first, and so we have a FIFO structure instead of a stack :

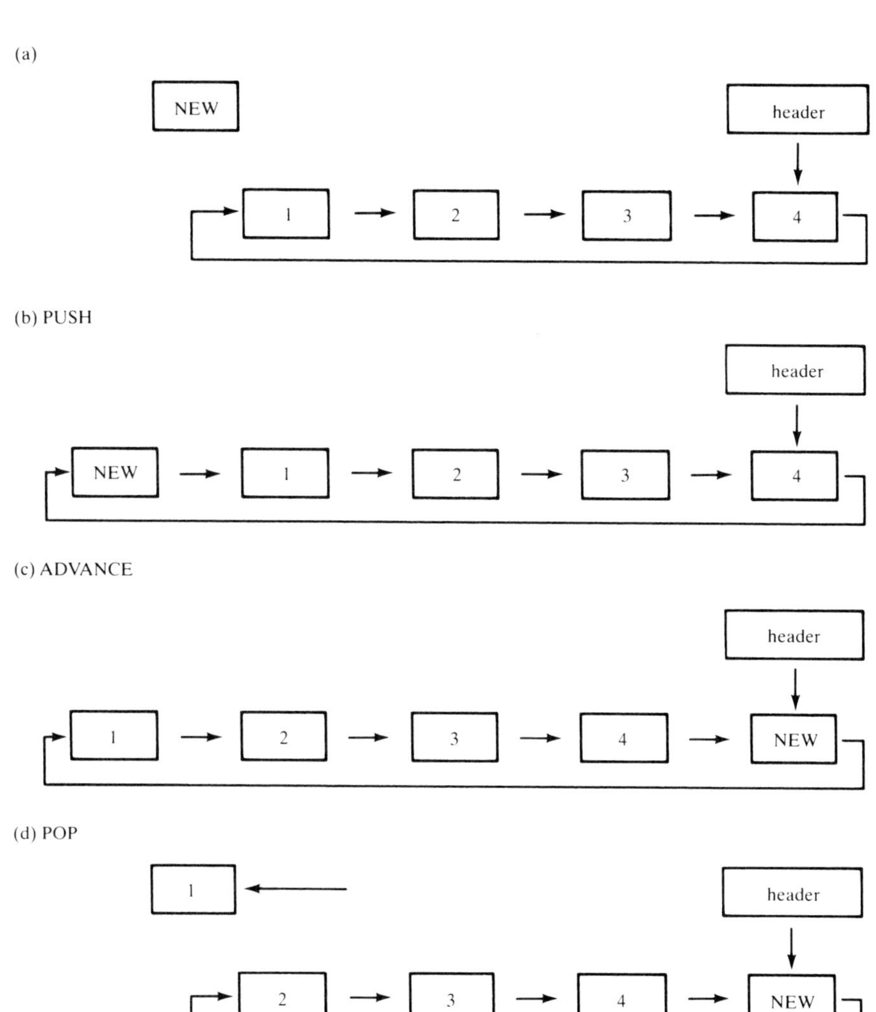
(a)
NEW
header
1
2
3
4
(b) PUSH
header
NEW
1
2
3
4
(c) ADVANCE
header
1
2
3
4
NEW
(d) POP
1
header
2
3
4
NEW

The word ADVANCE is hardly taxing to write, as it merely fetches the next pointer and stores it into the header :

```
: ADVANCE    DUP @ @ SWAP ! ;                  ( list ---   )
```

This now needs to be spliced into PUSH, which we can rename as ENQUEUE :

```
: ENQUEUE  GETNODE ?DUP                          ( value list ---   )
           IF  ROT OVER 2+ !
               OVER SWAP INSERT              -- keep a copy of "list"
               ADVANCE
           ELSE ." no free space " ABORT
           ENDIF ;
```

POP remains completely unchanged, though for consistency we should rename it DEQUEUE.

The circular list provides a very elegant way to implement queues. It gets by with only a single pointer to the list, whereas both head and tail pointers would be needed if a linear list were used. It does away with the modulo arithmetic required in our array implementation of the last chapter. It also has all the virtues of linked list implementations in general; any number of queues can share the same memory pool so long as the free list has space.

Circular queues are very useful in discrete simulations, for buffers of all kinds (e.g. a print spooler), and for scheduling in multitasking systems. The task queue can be most naturally represented by a circular list of task descriptors, which can be endlessly traversed to implement the popular "round robin" scheduling algorithm. In more sophisticated multitasked systems which employ semaphores to control access to resources, further circular buffers can be used to queue the tasks waiting for a particular resource.

The circular list has some other attractive properties that should be pointed out. In particular it is very easy to free the memory occupied by a circular list, and equally easy to concatenate two circular lists. In both cases the task can be performed by swapping two pointers, whereas for a linear list it would involve traversing the whole list. For example, our definition of KILL for a linear list requires the freeing of one node at a time. The time taken is thus proportional to the size of the list, and becomes large for long lists.

Concatenation of two circular lists can be accomplished by the simple swapping of pointers depicted below :

(a)

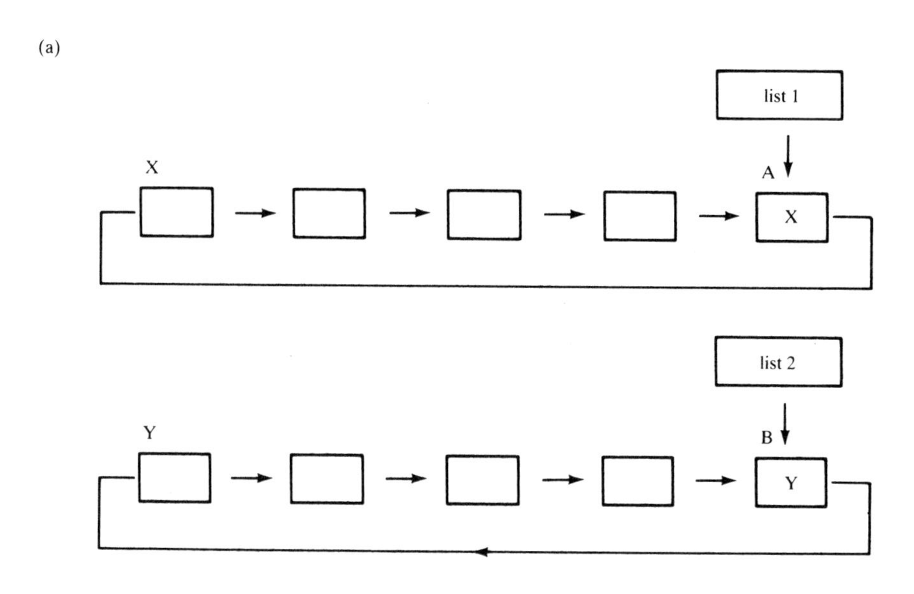

(b)

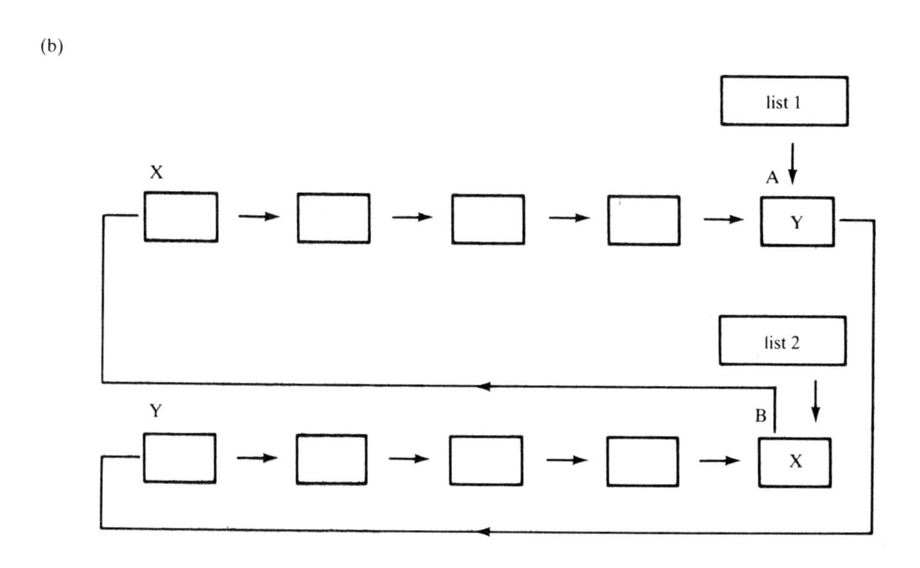

Merely swapping the nextnode pointers of the last nodes of the lists concatenates them into a single circular list. Using the terminology of the diagram, the operations are Y A ! and X B ! where A,B,X and Y are all addresses. The result is that both list1 and list2 now point to the same list, though to different parts of it.

To avoid having two pointers to the same list we could simultaneously set one of the list headers to zero, making it empty (the operation would then perhaps be better described as "decanting" rather than concatenation). Here is a Forth word to perform this procedure :

```
: CONCAT                                        ( list1 list2 ---    )
            DUP @ ?DUP                          --  is list2 empty?
            IF ROT DUP @ ?DUP                   --  is list1 empty?
               IF ROT DUP @ ROT DUP @           --  fetch X and Y
                  ROT ROT ! SWAP !              --  swap X with Y
                  DROP                          --  list1 not needed
               ELSE !                           --  list1 = list2
               ENDIF
               0 SWAP !                         --  make list2 empty
            ELSE 2DROP                          --  do nothing
            ENDIF ;
```

Note the complication introduced into our simple schema by the need to test for empty lists (again a consequence of the purely conventional nature of the zero header). The procedure can be summed up as "do nothing if list2 is empty, otherwise if list1 is empty just point it to list2, otherwise concatenate them. In all cases make list2 empty".

Using this definition, we can now trivially write a version of KILL for circular lists, since killing such a list is just equivalent to concatenating it with the free list (which is list1). Hence :

```
: KILL   FREELIST SWAP CONCAT ;
```

This KILL takes the same (minimal) time to free lists of any size.

The ease of the CONCAT operation suggests that circular linked lists will be useful in writing text processing programs. A linked list of characters would be a very flexible way to implement dynamic strings, but is too expensive in memory terms for most applications; a two-byte overhead for every one-byte character in the string is hard to stomach. It is quite feasible however to represent lines of text as circular lists of words (i.e. strings represented by normal byte arrays) and pages of text as circular lists of pointers to lines. A text editor based on such principles allows very fast "cutting" and "pasting" of text, using concatenation of lists; no data need be moved to make room for the insertions.

Addition of Large Numbers Using Circular Lists

Let us look at a completely different example of a use for the circular list.

Adding positive integers of arbitrary magnitude is not an operation which Forth normally supports. In writing routines to perform such arithmetic, one falls back on the "primary school" methods of adding (or subtracting etc.) one place at a time and passing any carries along in the direction of calculation.

Such an algorithm suggests that the linked list may be a good representation for the numbers. To demonstrate that this is so, we shall develop words to add together arbitrarily large positive integers, whose size is limited only by the available memory.

Conceptually the most simple method would be to represent each digit of a number as a list element, but this would be wasteful of memory. A Forth 16-bit single number can hold at most four digits in binary representation since 9999 can be held as a single 16-bit quantity, but 99999 cannot. We shall therefore "pack" four digits into into each list element. To start with we shall manually divide the numbers into four digit groups (later we can add a word to do this for us). So 5648329123 will be entered as 56 4832 9123.

Each of the numbers to be added will be represented as a circular queue of these four digit elements. The numbers will be placed on the parameter stack as list constants and then assigned using SET. This means that the least significant digits will be stored first :

```
{ 56 4832 9123 }  NUMBER1LIST SET
```

"Long" addition requires us to work from the least significant end of the numbers, and so we want the first in to be the first out; this is just what a queue gives us.

The addition will be performed by REMOVEing one element from each list and adding them together, then adding in any carry value from the previous addition, which is stored in a variable. The problem of dealing with two numbers of different lengths becomes trivial, because our implementation of REMOVE returns nil when either of the lists runs out of elements. This zero value can be used to set a flag and then just added to its corresponding element from the other list. Only when *both* removed values are zero does the addition terminate.

Here is how it works with 5648329123 plus 221234 :

```
1)    56 4832 9123   remove and add       9123
                     ------------------>          +       =    0357  carry 1
           22 1234   carry = 0            1234 + 0

2)         56 4832   remove and add       4832
                     ------------------>          +       =    4855  carry 0
                22   carry = 1              22 + 1

3)              56   remove and add         56
                     ------------------>          +       =      56  carry 0
               nil   carry = 0               0 + 0

4)             nil
                     terminate
               nil
```

The answer 5648550357 is left on the stack, in groups, but in reverse order :

0357 4855 56

and we shall need an output word to put it into a more respectable form.

This is a reasonably efficient way to proceed if the numbers to be added are of similar size. It is however rather wasteful if one number is much larger than the other (e.g. 3582875728459634923432949429349193499 + 2) since it involves completely traversing the list representing the larger number while essentially doing nothing.

These are the words which do the main work :

```
VARIABLE CARRY

VARIABLE END1   VARIABLE END2          -- flags for end of lists

: ADD4    +                            -- add two numbers        ( n1 n2 --- n)
          CARRY @ +                    -- add in carry
          10000 /MOD CARRY ! ;         -- calculate new carry

: VAL1    REMOVE DUP                                             ( list --- n)
          IF    2+ @                   -- if there is an element left, get it
          ELSE  TRUE END1 !            -- or set end flag
          ENDIF ;

: VAL2    REMOVE DUP                                             ( list --- n)
          IF    2+ @                   -- ditto
          ELSE  TRUE END2 !
          ENDIF ;

: ADD     0 CARRY !                                       ( list1 list2 --- n n ...)
          FALSE END1 !  FALSE END2 !
          BEGIN   OVER VAL1 OVER VAL2             -- get two elements
                  END1 @ END2 @ AND NOT           -- are both lists exhausted?
          WHILE   ADD4 ROT ROT                    -- put result below lists
          REPEAT  2DROP 2DROP ;
```

One could save a few bytes by combining VAL1 and VAL2 in a single definition and using a switch, but it scarcely seems worth the loss of clarity.

Before we can use these words we shall need to rewrite the list assignment word SET to work with a queue rather than a stack style list, by replacing PUSH with ENQUEUE :

```
: SET   DUP KILL                      -- empty the list ( ...n c list --- )
        SWAP 0 DO  SWAP OVER ENQUEUE  -- enqueue the items
               LOOP  DROP ;
```

We could now add two numbers, as in the example above, by :

```
NEWLIST A  NEWLIST B

{ 56 4832 9123 }  A SET

{ 22 1234 }  B SET

A B ADD
```

Notice that the algorithm we have used destroys both lists A and B, that is it leaves them empty. An algorithm which preserves the original lists is not difficult but requires that a non-destructive alternative to REMOVE be written. The result of the addition is left in the rather inconvenient form mentioned above, i.e. in reverse order on the stack. To print it out as a single string of digits we need to use the Forth formatted output words <# and #>. In order to do this we need to know how many digit groups there are in the answer. There are various ways to achieve this, but an easy one which is already to hand is to use our list delimiting words { and }, which work by counting the items left on the stack between them. So instead of saying A B ADD, let us say { A B ADD } which will leave a count of the items on top of the stack.

The required formatted output word looks like this :

```
: BIG.    SWAP S->D <# #S #> TYPE                 -- the first item might be short
         1- ?DUP IF 0 DO
                   S->D <# # # # # #> TYPE    -- all the rest are 4 digits
                 LOOP
              ENDIF ;
```

Once again a non-standard word, S->D, crops up. This word converts a single number to a double number with sign extension, and is much to be preferred to the sloppy practice of just pushing a 0 (even though in this case we are only concerned with positive integers). For those who do not have it, it is defined :

```
: S->D   DUP 0< IF -1 ELSE 0 ENDIF ;                ( n --- d )
```

(in Forth-83 it can be reduced to : S->D DUP 0< ; which is naughty but neat).

Now at least we can see the answers in a readable form :

```
{ 56 4832 9123 }  A SET

{ 22 1234 }  B SET

{ A B ADD }

BIG. 5648550357
```

However it would be much better if we could *enter* the numbers as continuous digit strings too; it is quite hard to divide large numbers by eye, especially when typing them from the "wrong" end.

It turns out that the code to accomplish this is actually rather larger and harder to write than the arithmetic routines themselves; a paradox typical of Forth. The reason for this is that the whole number must be read in, so that its length is known, before it can be divided into four digit groups. This entails the use of WORD as the input agent, and leaves the number as a continuous string (which incidentally is thereby limited to 255 digits). This must be sliced into four digit chunks which can be turned back into single precision numbers by CONVERT. Unfortunately CONVERT requires a counted string terminated by a blank or zero as its argument, and so we cannot simply scan along the string left by WORD; instead we have to move each slice to a new location (the PAD being the obvious place) before conversion.

The code looks like this :

```
: SLICE  DUP >R  DUP PAD C!              -- set count byte ( addr count  --- n)
         PAD 1+ SWAP CMOVE               -- move slice to PAD
         0 0 PAD CONVERT                 -- convert to number
         PAD R> +                        -- address of end of slice
         < IF ." illegal number" ABORT  -- check that it was all converted
           ENDIF DROP ;

: BIG    PAD 6 BL FILL                   -- make background of blanks
         BL WORD COUNT 4 /MOD            -- read in number, see how many groups
         SWAP ROT 2DUP + >R              -- stash addr of first four digit group
         SWAP ?DUP IF   SLICE SWAP       -- deal with any short first group
                  ELSE DROP
                  ENDIF
         R> SWAP
         ?DUP IF 0 DO DUP                    -- if there are any four digit groups
                      I 4 * + 4 SLICE SWAP  -- deal with them
                  LOOP  DROP
              ENDIF ;
```

BIG reads in the following big number and leaves it on the stack as a sequence of single numbers; hence it must be used inside list brackets :

```
{ BIG 5648329123 }  A SET

{ BIG 221234 }  B SET

{ A B ADD }

BIG. 5648550357
```

As a final syntactic gloss, we could create some definitions such as

```
: A<  { BIG ;      : >A  } A SET ;     : B<  { BIG ;     : >B  } B SET ;

: =>  { A B ADD } BIG. ;
```

to save a few keystrokes :

```
A< 5648329123 >A    B< 221234 >B  =>  5648550357
```

This offers an opportunity to raise a rather fine point of Forth programming style. Note that I have resisted the temptation to create an appearance of infix arithmetic (say by defining : BIG+ } A SET { BIG ;). It is very easy in Forth to bend the surface syntax in this way, so that the names of the words no longer reflect at all what they do. This is very poor style and can cause endless problems for those unfortunates who have to maintain or modify your code.

Extending the above scheme to cope with subtraction is not too difficult, but multiplication and division require rather more complex algorithms, and will tend to be very slow. Those readers who wish to try are advised to start with a simpler "unpacked" representation in which one list node per digit is used. It is also helpful to alter the circular list implementation slightly so that it maintains a count of the list elements in the list header, since multiplication and division will demand more than one pass over the lists.

4 Memory Management Using a Heap

In the last chapter we saw a mechanism for managing memory objects of fixed size, namely list nodes. "Managing" implies the ability to create and use objects and then to destroy them and reclaim the memory they occupied. A free list was used to manage allocation from a pool of list nodes.

The generalization of this kind of memory management is a system which can manage memory objects of varying sizes, and the data structure used for such memory management is often called a "heap" (suggesting something less ordered than a stack).

A heap consists of a pool of memory from which chunks of any size can be called off and used. When finished with, these chunks can be returned to the pool. A heap could be used for example to implement arrays whose size can be varied at run-time (unlike static Forth arrays whose size is fixed when they are compiled into the dictionary). The heap is also a useful way to store text strings, which are typically objects of varying size.

The principal programming problem in implementing a heap lies in avoiding or curing memory "fragmentation". We can depict a typical heap like this :

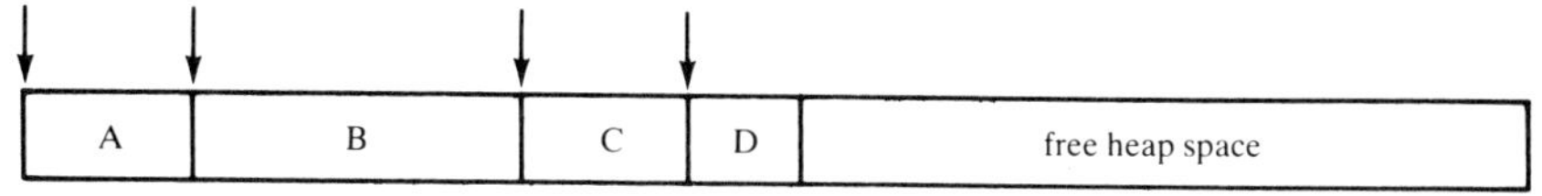

where A, B, C and D are pieces of memory currently in use by a program. If object B is no longer needed and its memory is returned to the heap, the situation will look like this :

A	free space	C	D	free heap space

The free space is now fragmented into two separate areas. It should be clear that as the heap is used more and more, the free memory will become increasingly fragmented. In particular, when a new piece of a certain size is requested for use, there will be no way of knowing where to find a piece large enough; worse still there may not **be** a piece large enough, even when most of the heap space is not in use.

Forming a linked list of the free spaces, as we did for list nodes, is not a sensible solution here. Since list nodes are all the same size, we can with confidence take the first node from the free list whenever a new node is required. With a heap however, every piece on the list could be a different size. Each piece would need to have a size field, and requesting a new piece would involve a time-consuming search of the whole list looking for a piece big enough; even then it might fail to find one.

A heap manager needs to be able to **compact** the heap, to restore all the free space to a contiguous block :

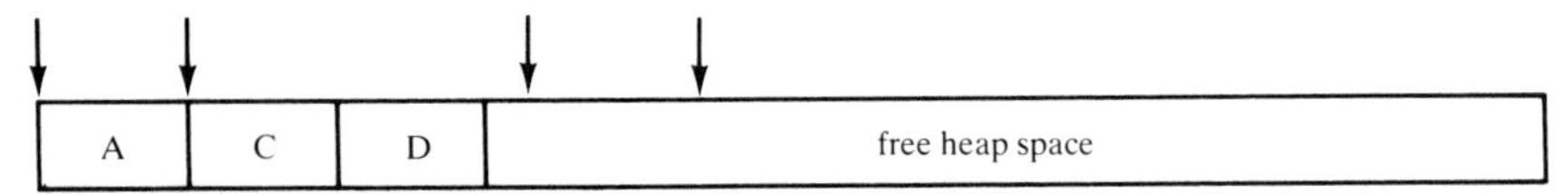

This guarantees that, assuming there is enough total heap space free to satisfy a request, then a large enough piece will be found. Compaction poses no physical problems in Forth, which supports fast block memory moves using MOVE and CMOVE. The heap can be compacted merely by using CMOVE to move all the heap items above the deleted one down to fill the gap.

However compaction does introduce a further complication. In the above diagram, objects C and D have been moved by the compaction process; in other words their addresses are no longer the same. Any pointers to these objects are now pointing to the wrong place (shown by the arrows). So in addition to compacting the heap, it is necessary to adjust all pointers to objects on the heap which have been moved. This could be a nightmare task if a program contains numerous references to heap objects.

Heap compaction with pointer adjustment is made feasible by using **indirect** references to the heap. Rather than giving out the actual address of a memory block in the heap, we shall instead give out the address of a "handle" which **contains** the address of the block. These handles, which are just 16-bit storage slots, will be kept in a table and pointer adjustment carried out on the contents of the handles. Programs which use the heap will work with the address of the handle itself, which never changes, and so pointer adjustment will be transparent to such programs :

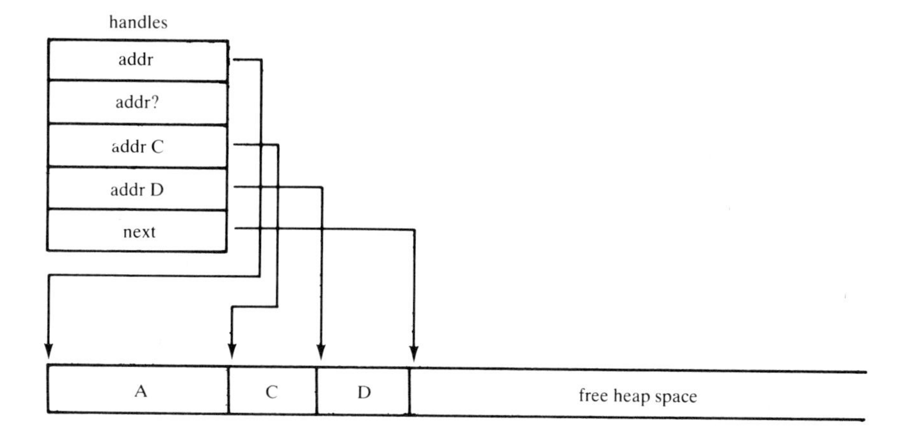

How many handles will be needed? Of course that depends upon how many objects need to be stored on the heap, and cannot be known in advance. Therefore we must be able to create new handles as they are needed.

What happens to the handle when a piece of memory is given back to the heap (as B was above)? One thing is certain; the handle cannot be removed from the table. The whole reason for introducing handles is that their addresses never ever change once they have been created. This means that handles which point to reclaimed pieces of memory must stay where they are and be made "inactive" until they are needed again.

In fact, since handles are fixed length objects, it makes sense to manage them like list nodes. All the reclaimed handles can be linked together into a free list. When a new handle is required we shall first try to re-use one from this free list, and only if it is empty does a new handle need to be created.

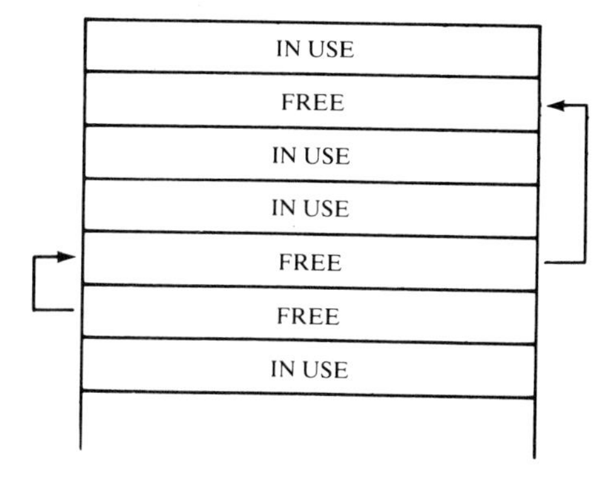

The disadvantage of such a scheme is that the total number of handles in existence can never diminish, but only increase or remain the same. It is possible to think of a pathological case where a huge number of heap objects (and hence handles) is created, then all but one are reclaimed; a lot of memory will be permanently wasted by the unused handles. This problem is unlikely to be encountered in real life so long as the heap space is large compared to the size of a handle. Heap usage will tend to fluctuate up and down in a typical application, and most handles will be reused many times.

It is theoretically possible to reclaim some of the space occupied by unused handles; only those at the end of the table (i.e. with no active ones following them) can be so reclaimed. However, by using a free list we forfeit any knowledge of the order in which handles will be used, and so the only way to identify such reclaimable handles would be by brute search.

The Implementation

In this Forth heap implementation a fixed chunk of dictionary space will be devoted to heap and handle storage (I am endebted to Bill Dress [DRESS85] for the basic idea). Handles are allocated from one end (high addresses growing downwards) while heap objects are allocated from the other end (low addresses growing upward). This is reminiscent of the way the stack and dictionary grow toward one another in the Forth compiler itself. As in Forth, we shall test for collision between a new heap object and the handle table, which signifies that the heap is full.

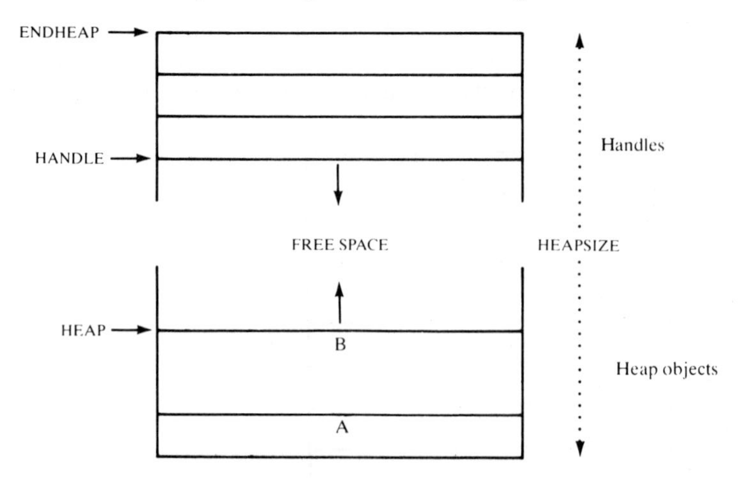

As shown in the diagram, three important pointers are maintained; one fixed on to the end of the whole heap area, and one each to the address of the next handle and heap object.

The size of a heap object will be stored in the object's first two bytes, and the address pointed to by the handle will be the first one after this size field :

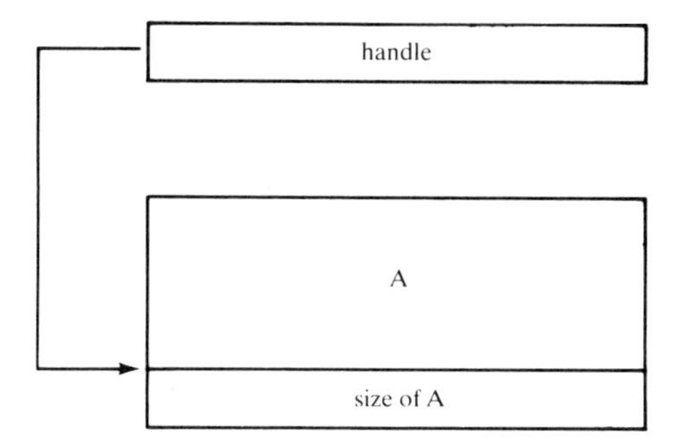

Other schemes are possible. It is quite an attractive idea to store the size in the handle itself (giving four byte handles); this would save an extra fetch operation when getting the size of an object. However, this causes problems with zero sized objects, for if two or more consecutive objects had zero size, their handles would all point to the same address. Choosing to put the size in the object means that even a zero sized object will have a physical size of two bytes and so the size fields double up as separators. Of course we might choose to forbid zero sized objects, but this would be to lose generality; the null string for example is a zero sized object.

Only two basic words are required to manage the heap. The first, called ALLOC takes a size from the stack, and returns the address of a handle to a piece of memory in the heap of that size. If the allocation fails due to shortage of heap space, then ALLOC will abort the program.

FREE takes a handle address from the stack and removes the piece of memory that it points to from the heap, by compacting the heap. The handle is returned to the free list to be reused, and all handles which point to heap objects that were moved in the compaction are adjusted to point to their correct new addresses. These objects are easily recognized; they are all those whose addresses are higher than that of the removed object.

Later on we shall develop a third word, RESIZE, which alters the size of a heap object while preserving its contents. This is useful for resizing arrays allocated in the heap.

In the following code, 'hdl' is used in stack annotations to mean the 16- bit address of a handle.

```
-- Heap manager

    16000 CONSTANT HEAPSIZE
        0 CONSTANT NIL

    CREATE HEAP   HEAPSIZE  ALLOT                    -- Create the heap

    HERE CONSTANT ENDHEAP>                           -- Top of heap area
```

```
    VARIABLE HEAP.PTR                                    -- Heap pointer
    VARIABLE HANDLE.PTR                                  -- Handles pointer
    VARIABLE FREE.HANDLE                                 -- Free list pointer

-- Very drastic; make sure you mean it!
    : HEAP.RESET    HEAP HEAPSIZE 0 FILL                 -- Initialize to zeroes
                    HEAP HEAP.PTR !                      -- Reset all pointers
                    ENDHEAP> 2- HANDLE.PTR !
                    NIL FREE.HANDLE ! ;

    HEAP.RESET                                           -- We mean it here

    : HEAP>      HEAP.PTR @ ;                            -- Fetch heap pointer
    : +HEAP>     HEAP.PTR +! ;                ( n --- ) -- Advance heap pointer

    : HANDLE>    HANDLE.PTR @ ;                          -- Fetch handle pointer
    : +HANDLE>   HANDLE.PTR +! ;              ( n --- ) -- Advance handle pointer

-- Adjust by 'n' the contents of handles which point to objects above 'addr'.
    : ADJUST.HANDLES   SWAP NEGATE                       ( n addr --- )
                       ENDHEAP> HANDLE> 2+               -- Range of handle space
                       DO I @ 2 PICK >                   -- {see below}
                          HANDLE> I @ - 0>  AND
                          IF I OVER SWAP +! ENDIF
                     2 +LOOP 2DROP ;

-- Return a new or second-hand handle.
    : GET.HANDLE    FREE.HANDLE @ ?DUP 0=                 ( --- hdl)
                    IF     HANDLE>  -2 +HANDLE>          -- Make new handle
                    ELSE   DUP @ FREE.HANDLE !           -- Reuse old handle
                    ENDIF ;

-- Will handle table collide with heap if 'size' bytes are allocated?
    : ?FULL   DUP HANDLE> HEAP>  ROT + 2+ - 0<           ( size --- size)
              IF ." No more heap space" CR ABORT ENDIF ;

-- Check that specified size is not negative.            ( size --- size)
    : ?NEG    DUP 0< IF ." Negative allocation" CR ABORT
                   ENDIF ;

-- Allocate space on the heap for an object.
    : ALLOC   ?NEG ?FULL                                 ( size --- hdl or 0)
              GET.HANDLE                                 -- Get a handle
              HEAP> 2+ OVER !                            -- Put pointer in handle
              OVER HEAP> !                               -- Put size into object
              SWAP 2+ +HEAP> ;                           -- Bump heap pointer

-- Get size of object
    : SIZE?    @ 2- @ ;                                  ( hdl --- size)

-- Return a handle to the free list.
    : RELEASE.HANDLE   FREE.HANDLE @ OVER !              ( hdl --- )
                       FREE.HANDLE ! ;

-- Reclaim the memory occupied by an object.
    : FREE    DUP SIZE? 2+                               ( hdl --- )
              OVER @ 2-                                  -- Source for move
              2DUP + DUP >R                              -- Destination for move
              SWAP HEAP> R@ - CMOVE                      -- Compact heap
              DUP NEGATE +HEAP>                          -- Adjust heap pointer
              SWAP RELEASE.HANDLE                        -- Free the handle
              R> ADJUST.HANDLES ;                        -- Adjust handles
```

Note that the two test words ?NEG and ?FULL, which check for legal parameters, both cause an ABORT if they fail. For many applications this is quite sufficient. However in some circumstances this behaviour may be too brutal, and it may be preferable to have ALLOC return a zero flag instead of a handle should the allocation fail due to lack of space. The application could then test this flag and perhaps perform some garbage collection to free up space rather than aborting. ?NEG however should always cause a direct ABORT, since a negative allocation will irretrievably corrupt the heap.

The word ADJUST.HANDLES requires rather more explanation than could be fitted into comments. It performs a run through the whole of the handle table (between addresses ENDHEAP> and HANDLE>) looking for handles which point to objects that have addresses higher than that of the newly removed object. In addition however it has to reject all handles which are not in current use. This it can do because they are linked together into a free list; in other words they contain either nil or an address which lies within the handle table itself (i.e. is greater than HANDLE>). Those handles which are selected are adjusted by a fixed offset which is simply the distance that the heap was moved to compact it.

Using the heap is quite straightforward, though it requires some discipline when manipulating handles. To create an object on the heap we could say :

```
VARIABLE FRED
25 ALLOC FRED !
```

FRED now contains a handle to a newly allocated 25-byte piece of heap space. The size of this piece can be found by FRED @ SIZE? and its address by FRED @ @. Even though the code initializes the heap to all zeroes, for neatness, you cannot rely on this after the heap has been in use for some time; it will fill up with garbage, and so new pieces ought to be explicitly initialized.

It is important to remember that if the handle is lost then that object can never be accessed again and its memory cannot be reclaimed. The handle can be lost quite easily, by merely storing something else in FRED, for example,

```
16 ALLOC FRED !
```

FRED now contains a new handle, and the old one is irretrievably lost. What we should have done is to release the old object **before** assigning anything new to FRED :

```
FRED @ FREE     16 ALLOC FRED !
```

This is one kind of discipline. The other is to remember that the whole heap can quite easily be corrupted by performing illegal operations on handles. For example storing a value directly into a handle is disastrous, as also is trying to use a handle which has been released. After FRED @ FREE, FRED still contains the address of the freed handle, which could be misused. FRED should be reinitialized to zero after a FREE, unless it is reassigned to immediately.

All these cautions suggest that handles are highly suitable recipients for the protection offered by an abstract data type. Using the system developed in Chapter 2, we could define a type HEAPVAR to use instead of an ordinary Forth VARIABLE. The permitted operations on this type will safeguard the integrity of the heap by removing the need for (and indeed preventing) the direct manipulation of handles :

```
TYPE> HEAPVAR
2 VAR HANDLE
OPS>
: INIT    0 ALLOC HANDLE ! ;                 --  Create a new handle
: <<      HANDLE @ FREE  ALLOC  HANDLE ! ;   --  Allocate a piece      ( n ---   )
: >>      HANDLE @ @ ;                       --  Get address of piece (   --- n )
: <<?     HANDLE @ SIZE? ;                   --  Get size of piece    (   --- n )
ENDTYPE> HEAPVAR
```

These new operators we have invented, <<, >>, and <<?, allow all the desirable operations to be performed on a piece of heap memory, while preventing any of the errors outlined above. The new type can be used like this :

```
HEAPVAR FRED

25 FRED <<          -- Assign 25 bytes of space to FRED.

FRED >>             -- Put address of this space on the stack.

50 FRED <<          -- Double the space allotted to FRED.

FRED <<? .    50 ok
```

Note that the << operator automatically frees any existing allocation before assigning new space, so the loss of a handle becomes impossible. This obviates the need to use FREE directly.

A HEAPVAR will in fact keep the same handle for its whole lifespan, the one which is given to it by the INIT operation when it is created, because when the handle is freed in <<, it is immediately taken from the free list and reused by ALLOC. This however is an irrelevant aside; this level of detail must no longer concern us. As long as HEAPVARs are used, we should never have to think about handles at all.

String Storage on the Heap

As an example of the use of a heap, we shall implement a simple system of dynamic storage of strings, similar to that found in many Basic interpreters.

Strings have always been a weak point in Forth because successive standards have failed to decide how to refer to them consistently. Forth uses both "packed" string references, i.e. the address of a string whose first byte contains the character count, and "unpacked" references, i.e. the address of the first character and the count. For example, WORD returns the address of a packed string, but TYPE wants an unpacked string argument, so the word COUNT must be used to convert one to the other. (Just for good measure Forth also throws in null terminated strings in the inner workings of EXPECT, which covers the full gamut of possibilities.)

Our basic word for placing strings on the heap is $>HEAP, which takes an unpacked string argument and returns a handle :

```
: $>HEAP  DUP ALLOC                          ( addr count --- hdl)
          DUP >R
          @ SWAP CMOVE R> ;                   -- Copy string
```

The word works very simply by allocating a piece of heap memory and then using CMOVE to copy the string text into it (the count is not copied as the heap already supplies one).

Using $>HEAP we can now define words to create string literals and to take strings from the input stream. We have already seen (in Chapter 2) the use of the non-standard word " to create string literals. However here it will be redefined to produce a handle rather than a packed string address :

```
: "      ASCII " WORD COUNT $>HEAP                      ( +++ hdl. )
         STATE @ IF [COMPILE] LITERAL ENDIF ;     IMMEDIATE

: IN$    BL WORD COUNT $>HEAP ;                          ( +++ hdl )
```

The word ASCII is not a standard word, but is a great aid to readability. It merely returns the ASCII value of the character which follows it. If your dialect does not have it just use 34 WORD COUNT (34 being the ASCII code for "). Notice that " has been defined as a state smart, immediate, early binding word. When used in a colon definition it will create a string on the heap at compile-time and compile its handle into the definition.

These definitions allow us to use ordinary Forth variables to hold strings, since handles are just 16-bit numbers. For example,

```
VARIABLE  FRED$
" MISSISSIPPI" FRED$ !

: NEWFRED  FRED$ @ FREE   IN$ FRED$ ! ;
NEWFRED AMAZON
```

Note the way that the old content of FRED$ is freed before reassigning to it, thus ensuring that the space occupied by MISSISSIPPI is reclaimed. Keep in mind that the content of such a variable is a handle, and it will require further processing after retrieval before it becomes a string again. A word >$ can be defined which turns a handle back into an unpacked string :

```
: >$     DUP @ SWAP SIZE? ;                      ( hdl --- addr count)
```

Now we could say, using the above examples,

```
FRED$ @ >$ TYPE
```

to print the string value held in the variable. For convenience we might even define

```
: $.     >$ TYPE ;                               ( hdl ---    )
```

allowing us to say FRED$ @ $. which is slightly neater. There is endless scope for playing with the syntax of string manipulation in this way, but one should always keep in mind the goals of consistency and security rather than mere prettification. We have taken one step towards consistency above, in that both string literals and variables return the same type of object, namely a handle. Avoiding mixed representations helps to forestall programming errors.

A useful string manipulation word to have would be one which concatenates two strings. Such a word, called $+, will take the handles of two strings as arguments and return a handle to a new string formed by joining them. It is non-destructive, so neither of the original strings is altered; the corollary of this is that the amount of heap space occupied by the strings will be doubled.

Here is a possible definition of $+ :

```
: $+      2DUP                                      ( hdl1 hdl2 --- hdl3)
          SIZE? SWAP SIZE? +                        -- Size of combined strings
          ALLOC                                     -- Allocate space
          ROT >$ DUP >R  2 PICK @ SWAP CMOVE         -- Copy first string
          SWAP >$  2 PICK @ R> + SWAP CMOVE ;        -- Copy second string
```

As an example of its use we could try :

```
VARIABLE A$   VARIABLE B$   VARIABLE C$

" MATTER" A$ !     " HORN" B$ !

A$ @ B$ @ $+   C$ !
```

The variable C$ now contains a handle to the new string MATTERHORN.

Using similar methods, all the other popular string functions, such as equality tests and substring extraction can be defined.

The alert reader will have noticed that in this code I have slipped back into the direct manipulation of handles, with all the risks that implies for the integrity of the heap. As a reward for such alertness, I leave it to you as an exercise to write an abstract data type representation for string handles. The way that handles are used in HEAPVAR above can serve as a starting point. The action of $>HEAP needs to be put into a type operation, and " can be returned to its earlier usage, namely to just put the address of a string onto the stack.

Resize

For some applications it will be very useful to have the ability to change the size of a piece of heap memory without losing its current contents. One example would be a program which allocates an array in the heap and needs to increase or decrease the size of the array at run-time.

The word RESIZE seems quite straightforward at first glance. It is merely an extension of the way that heap compaction is performed. The heap is moved up or down to create the correct new size for the designated piece, and then any affected handle pointers are adjusted as before. However, when we come to implement RESIZE we find that Forth's CMOVE is not adequate in these circumstances.

The block moves involved in heap management are very often overlapping moves, that is, the source and destination ranges for the move overlap. CMOVE cannot cope correctly with overlapping moves towards high memory. It works adequately for heap compaction because the heap is always moved downwards (due to the design decision to have the heap grow upwards). But with RESIZE, moves might have to go in either direction depending upon whether we are increasing or decreasing the size of a piece. Forth-83 includes a CMOVE> word which can do correct overlapping moves to high memory, and so we could test the direction of the move in an IF, and use CMOVE or CMOVE> as appropriate.

There is something else worthy of consideration in the design of RESIZE. If a piece of memory is the top piece on the heap, then resizing it becomes trivial (a matter of simply adjusting the heap pointer, with no data movement at all). This case should certainly be tested for as it will result in a considerable speed increase. But this case also suggests a possible alternative strategy for RESIZE. Instead of resizing a piece in its original position (which involves choosing between CMOVE and CMOVE>), why not copy the piece to the top of the heap, perform the trivial resize, and then reclaim the original using FREE which we have already written? This strategy has an additional benefit. If the same piece is being repeatedly and consecutively resized (as might happen with an important array), then this method optimizes the operation. The first resize will be relatively slow, but all subsequent ones will be as fast as possible since the piece is now top of the heap.

Here is the code for RESIZE using the second strategy :

```
: RESIZE  ?NEG                              ( hdl size ---  )
          OVER @ 2-                          -- Get addr of size field
          2DUP @ -  2-                       -- Calculate size change
          ?FULL DROP                        -- Enough room?
          DUP DUP @ + 2+ HEAP> =             -- Is it top of heap?
          IF    2DUP @ - +HEAP>              -- Adjust heap pointer
                ! DROP                      -- Store new size
          ELSE HEAP> OVER @ 2+ CMOVE         -- Copy to heap top
                ALLOC SWAP FREE             -- Allocate new space; free old
                DUP @ GET.HANDLE !          -- Restore original..
                RELEASE.HANDLE              -- ...handle
          ENDIF ;
```

RESIZE is used like this :

```
VARIABLE FRED
25 ALLOC FRED !
FRED @ 30 RESIZE
```

which leaves FRED pointing to a 30-byte piece.

Using RESIZE we could define a destructive version of $+ which places the result of the concatenation into one of the original strings, thus tying up less heap space. To do this we would RESIZE the target string to the combined string size before copying the other into it, rather than ALLOCating the combined size as was done above.

Objects and the Heap

The heap is a very powerful data structure indeed. Many high level languages such as Pascal and C provide a heap as one of the resources available to the programmer. It is particularly useful in the sort of advanced (often interpreted) languages used for Artificial Intelligence research, where there is a requirement for highly dynamic data structures which are created and destroyed "on the fly". Heaps are also used in most operating systems to manage the memory and apportion it to requesting processes.

The above code shows that a simple and efficient heap structure can be written in high-level Forth. However in traditional 16-bit Forths there is typically very little memory available to implement a reasonably sized heap. The above example (16000 bytes of space) filled almost half the free memory on the author's system. The heap is likely to become of much greater importance to Forth programmers once 32-bit implementations become the norm, and dictionary space is measured in megabytes. In these circumstances it should become as integral a part of the system as the stacks and dictionary are at present.

The heap can be very effectively combined with the structures discussed in previous chapters. For instance, the combination of heap objects using linked lists of handles allows dynamic Lisp style list processing to be performed. List nodes could be drawn from the heap, but it is probably better to manage them in separate memory space via a free list as before; the data fields of active list nodes would then contain handles pointing into the heap.

We can quite easily modify our abstract data type mechanism (Chapter 2) so that objects are created on the heap instead of in the dictionary. Instead of using ALLOT to allocate space for a newly created object, we shall use ALLOC to create the object's body on the heap, and then compile the returned handle into its dictionary header. An object thus consists of a dictionary header containing a handle pointing to instance variables kept on the heap. At the minimum, only the word MAKE.INSTANCE needs changing :

```
--Create a new instance on the heap                    ( addr ---   )
: MAKE.INSTANCE  CREATE DUP @   ,            -- Store key into header
                        SZ@ ALLOC            -- Allot its storage on heap
                        ,                   -- Compile handle into header
                        IMMEDIATE
                 DOES>  DUP 2+ @ @ 2-          -- Get address of heap piece less 2
                        SWAP @               -- Get key
                        DO.OR.COMP ;
```

The 2- in this definition recognizes the fact that object bodies now no longer have a key field to skip over. Serious users will prefer to tweak the rest of the code so that this offset is not added in the first place (remove 2+ from OFFSET, add a 2+ to MAKE.INSTVAR). I leave as an exercise for the reader the conversion of our later INITIALIZEd version of MAKE. INSTANCE.

A new word, say ZAP, can now be written to free the heap space occupied by redundant objects (though their headers will remain in the dictionary), something like this :

```
: ZAP   ' >BODY                        -- get object's PFA
        DEAD @ OVER !                  -- scrub its key
        2+ DUP @ FREE                  -- free its heap space
        0 SWAP ! ;                     -- scrub handle too, for luck
```

Used as in ZAP FRED. The variable DEAD contains the key address of a specially created type called DEAD :

```
TYPE> DEAD
OPS>
ENDTYPE> DEAD
```

which ensures that such dead objects cannot be used to do any harm! A further word might be written to enable such "husks" to be reused later. At the cost of some loss of efficiency, we can now create and destroy objects "on the fly", a particularly useful ability in complex simulation programs. Obviously one would want to go further and modify MAKE.ARRAY to allocate its space on the heap, enabling the alteration of the size of object arrays at run-time by using RESIZE.

Heapforth??

Finally, one intriguing prospect is the introduction of the heap into the heart of the Forth compiler itself. Several authors have suggested similar schemes, and there are a few current implementations using token threading and separated headers which approximate this scheme. The basic idea is to separate the headers from the bodies of Forth words, and to allocate the bodies on a heap instead of in the singly linked list of the dictionary. The headers then become in effect handles. For the scheme to work properly, headers should be of constant size, which would mean placing some restriction on the storage of name fields :

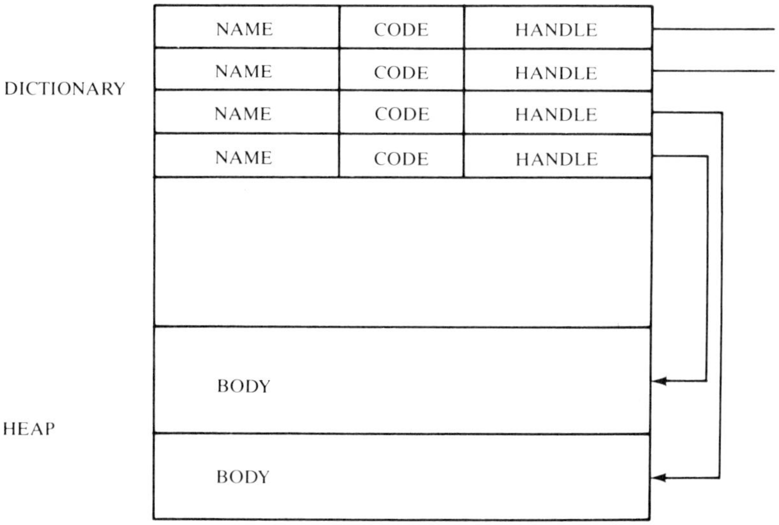

DICTIONARY	NAME	CODE	HANDLE
	NAME	CODE	HANDLE
	NAME	CODE	HANDLE
	NAME	CODE	HANDLE
HEAP	BODY		
	BODY		

The consequences for Forth would be quite profound, though it might remain recognizable as the same language. Dictionary search would proceed very efficiently using a fixed offset in place of link fields. The bodies of words could be edited and recompiled or even deleted without affecting any other words in the dictionary, by resizing them. Since headers would be reused like handles, dictionary order would become unimportant (with profound implications for scoping!). It would even be possible to compile forward references to undefined words (as in Lisp) if the compiler were to create a special NOOP body for such words, which could be filled out later. Limiting the number of available header/handles to 64,000 would enable 16-bit compilation addresses to be retained even in a 32-bit implementation, which would in turn permit a multi-megabyte overall address spaces, but with a code density more or less the same as for current Forth.

All these changes would be brought together in a resident smart editor, in which editing, deletion and compilation of words is effected by pointing and clicking with a mouse, source code being regenerated from the compiled code by an integral smart decompiler. Top down program design would be directly supported by merely entering the name of a yet to be written word, whereupon a "stub" is automatically generated.

Reference

Dress, W.B. (1986). "A Forth Implementation of the Heap Data Structure". *Journal of Forth Applications and Research,* Vol 3 No 3, pg 39. Institute of Applied Forth Research, Rochester.

Postscript

The Utopian ramblings at the end of the last chapter were provoked by the perception that Forth is poised to enter an exciting and challenging phase of its history.

Traditionally Forth has been based upon a 16-bit wide stack and 16-bit compilation addresses, thus limiting the size of its dictionary to 64 Kbytes. This has not presented very great problems in practice because threaded Forth code is much more compact than the code produced by other compiled languages, and more compact even than the average Assembly language program, so reasonably large programs can be accommodated in such a space. Nevertheless, because it is so frequently used for control applications destined to be placed in ROM, space saving has become a major part of programming style for the typical Forth programmer.

As 32-bit microprocessors begin to come on to the market, and even personal computers can have memory measured in megabytes, there is a swelling body of opinion that feels that Forth must move with the times and adopt a 32-bit stack and addresses, capable of utilizing such megabyte address spaces. Forth is becoming accepted for use in new application areas such as Real-time Expert Systems, where very large programs need to be written. Moving to a full 32-bit implementation of Forth is something which must happen soon, but to do it in a way which keeps some compatibility with the past will require much thought and good sense on the part of the Standards Team. It will also entail a gradual change of emphasis on the part of programmers, as saving the last byte makes less and less sense compared to writing secure and manageable programs.

Exactly the same changes are occurring in the realm of execution speed. Forth's popularity has always rested upon its being the fastest *interactive* language (typically 10 or more times faster than interpreted Basic). However it is not fast when compared to native-code compiled Fortran, Pascal or C, and still less so when compared to Assembly language. This is because the threading mechanism (i.e. compilation of addresses rather than machine codes) imposes a run-time overhead due to the execution of the inner interpreter code. Hence Forth programmers have typically been as concerned with saving cycles as they have with saving bytes. Optimization by rewriting inner loops in Assembler is a familiar part of many professional Forth programmers routine.

All this is about to change. In 1985, almost simultaneously and on opposite sides of the Atlantic, the first true hardware implementations of Forth were announced. These "Forth Machines", one from the Novix Corporation in California and the other from MetaForth Ltd. in England, are computers dedicated to running Forth; they are hardware realizations of the "virtual Forth machine" which has hitherto been simulated in software on a host computer. In other words Forth is the Assembly language of such computers. They feature hardware stacks which are not situated in the main address space of the processor, and concurrent "threaders" which effectively reduce the overhead due to the threaded code to zero (or less!), executing nested Forth definitions as if they consisted of in-line machine code. The effect on performance is dramatic. Both machines can execute of the order of 8 to 10 Million instructions per second, which makes Forth programs execute faster than Fortran programs on a VAX 11/780. They are also small enough to be easily contained in a desk top computer.

The eventual effect of the introduction of these machines (and others which are sure to follow) will be that performance will become a less critical issue than it is among today's Forth programmers. In particular the use of Assembler for optimization will disappear since these machines cannot be programmed at any level lower than Forth.

Given these two complementary trends, I am convinced that the time is ripe for Forth programmers to pay greater attention to issues of program design and structure, and less to squeezing out the last drop of performance. It is with such a vision in mind that I wrote this book. While writing it I was acutely aware of, and embarassed by, every compromise that had to be made which might reduce program performance. But then I would take a deep breath and think of the not too distant future when 10 MIPS and a megabyte of memory will make a mockery of such concerns.

Index